Collins

Maths Frameworking

3rd edition

Kevin Evans, Keith Gordon,
Trevor Senior, Brian Speed,
Chris Pearce

Contents

How to use this book

Learning objectives

See what you are going to cover and what you should already know at the start of each chapter.

About this chapter

Find out the history of the maths you are going to learn and how it is used in real-life contexts.

Key words

The main terms used are listed at the start of each topic and highlighted in the text the first time they come up, helping you to master the terminology you need to express yourself fluently about maths. Definitions are provided in the glossary at the back of the book.

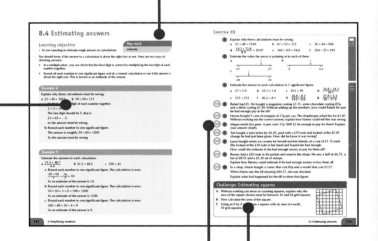

Worked examples

Understand the topic before you start the exercises, by reading the examples in blue boxes. These take you through how to answer a question step by step.

Skills focus

Practise your problem-solving, mathematical reasoning and financial skills.

Take it further

Stretch your thinking by working through the **Investigation**, **Problem solving**, **Challenge** and **Activity** sections. By tackling these you are working at a higher level.

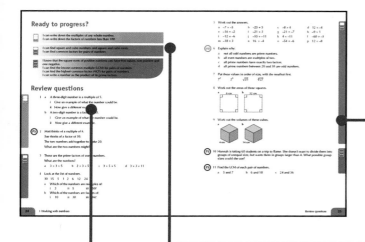

Progress indicators

Track your progress with indicators that show the difficulty level of each question.

Ready to progress?

Check whether you have achieved the expected level of progress in each chapter. The statements show you what you need to know and how you can improve.

Review questions

The review questions bring together what you've learnt in this and earlier chapters, helping you to develop your mathematical fluency.

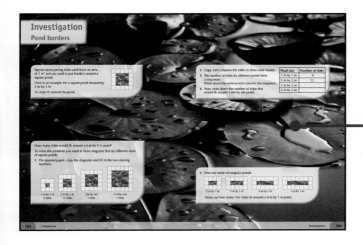

Activity pages

Put maths into context with these colourful pages showing real-world situations involving maths. You are practising your problem-solving, reasoning and financial skills.

Interactive book, digital resources and videos

A digital version of this Pupil Book is available, with interactive classroom and homework activities, assessments, worked examples and tools that have been specially developed to help you improve your maths skills. Also included are engaging video clips that explain essential concepts, and exciting real-life videos and images that bring to life the awe and wonder of maths.

Find out more at www.collins.co.uk/connect

1

Working with numbers

This chapter is going to show you:

- how to multiply and divide negative numbers
- how to find the highest common factor and the lowest common multiple of sets of numbers
- how to use powers and find roots
- how to find the prime factors of a number.

You should already know:

- how to add and subtract negative integers
- how to order operations, following the rules of BIDMAS
- what a factor is
- what a multiple is.

About this chapter

What games do you play? In many of them, you will certainly use numbers. For example, darts players need to make calculations very quickly. They must work out the target numbers they must score, to finish the game. Then they have to think about the possible combinations of scores from three darts.

Number skills are important in field games, such as rugby or cricket. They are essential in many board games and, of course, computer games. This chapter will help you develop those skills further for use in everyday life.

1.1 Adding and subtracting with negative numbers

Learning objective

- To carry out additions and subtractions involving negative numbers

You can use a number line to add and subtract **positive numbers** and **negative numbers**.

Example 1

Use a number line to work out the answers.

a $3 + (-7)$ **b** $(-5) + 3$ **c** $(-2) + (-5)$

a Starting at zero and 'jumping' along the number line to 3 and then back 7 gives an answer of –4.

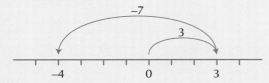

b Similarly, $(-5) + 3 = -2$.

Notice that you can use **brackets** so that you do not mistake the negative sign for a subtraction sign.

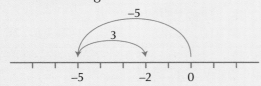

c Similarly $(-2) + (-5) = -7$.

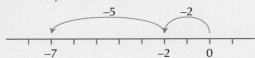

Remember, adding a negative number gives the same result as subtracting a positive number.

Example 2

Work out the answers.

a $5 + (-2)$ **b** $12 + (-4)$ **c** $(-4) + (-6)$

a $5 + -2 = 5 - 2$
          $= 3$

b $12 + -4 = 12 - 4$
          $= 8$

c $(-4) + (-6) = (-4) - 6$
            $= -10$

Remember that subtracting a negative number is the same as adding a positive number.

$4 - (-1) = 5$ has the same value as $4 + 1 = 5$ and

$4 - (-2) = 6$ has the same value as $4 + 2 = 6$.

Example 3

Work out the answers.

a $8 - (-3)$ **b** $10 - (-4)$ **c** $-9 - (-7)$

  a $8 - (-3) = 8 + 3 = 11$ **b** $10 - (-4) = 10 + 4 = 14$ **c** $-9 - (-7) = -9 + 7 = -2$

Exercise 1A

1   Use the number line below to help you work out the answers.

     a $3 - 5$ **b** $7 + (-4)$ **c** $6 + (-5)$ **d** $8 + (-4)$

     e $(-4) + (-4)$ **f** $6 - 10$ **g** $(-4) + 10$ **h** $1 - 9$

     i $11 + (-5)$ **j** $(-3) + (-7)$ **k** $13 + (-10)$ **l** $6 + (-6)$

     m $8 + (-9)$ **n** $11 + (-21)$ **o** $1 + (-5)$ **p** $(-3) + (-10)$

$-15\ -14\ -13\ -12\ -11\ -10\ -9\ -8\ -7\ -6\ -5\ -4\ -3\ -2\ -1\ 0\ 1\ 2\ 3\ 4\ 5\ 6\ 7\ 8\ 9\ 10\ 11\ 12\ 13\ 14\ 15$

2   Work out the answers.

     a $9 - 5$ **b** $6 - 4$ **c** $3 - 9$ **d** $5 + 9$

     e $3 - 6$ **f** $7 - 8$ **g** $-3 + 7$ **h** $-4 - 3$

     i $-6 - 2$ **j** $-3 + 10$ **k** $-6 - 6$ **l** $-11 + 9$

     m $-5 + 5$ **n** $-2 + 2$ **o** $-9 - 5$ **p** $-12 + 13$

3   Work these out.

     a $11 + (-6)$ **b** $(-6) + (-11)$ **c** $50 - 60$ **d** $(-11) + (-11)$

     e $30 + (-20)$ **f** $(-60) + 60$ **g** $15 + (-20)$ **h** $200 - 300$

     i $15 + (-8)$ **j** $(-200) + (-80)$ **k** $10 + (-40)$ **l** $(-7) + (-17)$

4   Copy and complete each calculation.

     a $9 - (-3)$ **b** $11 - (-6)$ **c** $21 - (-11)$ **d** $-11 - (-4)$ **e** $(-7) - (-7)$

       $= 9 + 3$ $= 11 + 6$ $= 21 +$ $=$

       $=$ $=$ $=$

5   Use the number line below to help work out the answers.

     a $5 - (-7)$ **b** $9 - (-2)$ **c** $(-6) - (-6)$ **d** $11 - (-3)$

     e $(-5) - (-3)$ **f** $(-2) - (-9)$ **g** $(-6) - (-9)$ **h** $(-11) - (-4)$

     i $9 - (-5)$ **j** $(-10) - (-4)$ **k** $(-8) - (-8)$ **l** $(-6) - (-16)$

     m $(-9) - (-3)$ **n** $(-20) - (-30)$ **o** $0 - (-20)$ **p** $(-3) - (-7)$

$-15\ -14\ -13\ -12\ -11\ -10\ -9\ -8\ -7\ -6\ -5\ -4\ -3\ -2\ -1\ 0\ 1\ 2\ 3\ 4\ 5\ 6\ 7\ 8\ 9\ 10\ 11\ 12\ 13\ 14\ 15$

6   Work these out.

     a $12 - (-9)$ **b** $(-11) - (-11)$ **c** $21 - (-12)$ **d** $(-16) - (-6)$

     e $22 - (-17)$ **f** $41 - (-21)$ **g** $-15 - (-10)$ **h** $-100 - (-250)$

     i $19 - (-8)$ **j** $(-250) - (-50)$ **k** $13 - (-23)$ **l** $(-5) - (-13)$

7   Copy each statement and work out the missing numbers.

     a $9 + \ldots = 5$ **b** $10 - \ldots = 15$ **c** $10 + \ldots = 3$ **d** $11 - \ldots = 17$

     e $\ldots + (-6) = 10$ **f** $(-4) - \ldots = 8$ **g** $\ldots + (-8) = 12$ **h** $(-5) - \ldots = 9$

8 Work out the answers.

a 7 − −3　　b −4 − −4　　c 9 − −6　　d −7 + −3　　e 10 − −3

f −9 − −7　　g −6 + −6　　h 11 − −4　　i −3 + −3　　j −4 + −10

k 5 − −6　　l 7 − −10　　m −2 − −3　　n − 3 + 3　　o −6 + −9

9 These temperatures were recorded at Manchester Airport in February.

Copy and complete the table.

Draw a number line to check your answers.

Day	Sun	Mon	Tue	Wed	Thu	Fri	Sat
Maximum temperature (°C)	6	1	−1		2	3	5
Minimum temperature (°C)	−5	−6		−7	−4		−2
Difference (degrees)	11		8	10		6	

 10 A fish is 6 m below the surface of the water.

A cormorant is 15 m above the surface of the water.

How many metres must the bird descend to get the fish?

Activity: Brick walls

A In these 'walls', subtract the number in the right-hand brick from the number in the left-hand brick to find the number in the brick below.

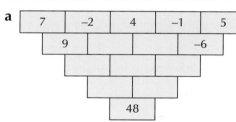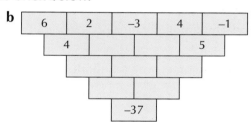

B Make up a wall as above that ends with:　**a** 50 and　　**b** −50.

1.2 Multiplying and dividing negative numbers

Learning objective

- To carry out multiplications and divisions involving negative numbers

This diagram shows the result of multiplying both positive and negative numbers by a positive number.
In this example all numbers are multiplied by +2.

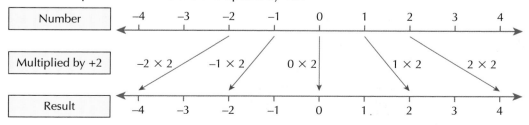

This shows that:

- multiplying a positive number by another positive number gives a positive number
- multiplying a negative number by a positive mumber gives a negative number.

You can summarise this as $(+) \times (+) = (+)$ and $(-) \times (+) = (-)$.

Example 4

Work out the answers.

a -2×4 **b** 3×-5

   a $-2 \times 4 = -8$ **b** $3 \times -5 = -15$

What happens if you multiply a number by a negative number?

This diagram shows positive and negative numbers multiplied by -2.

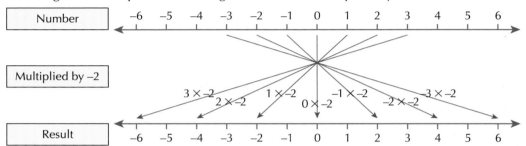

This diagram shows that multiplying a positive number by a negative number gives a negative result, as in the first diagram.

Here it is just shown the other way round. But this diagram also shows that multiplying a negative number by a negative number gives a positive number.

You can summarise this as $(-) \times (-) = (+)$ and $(+) \times (-) = (-)$.

To help you remember . . . .

- When multiplying numbers with different signs, the answer is negative.
- When multiplying numbers with the same sign, the answer is positive.

Example 5

Work out the answers.

a $4 \times (-5)$ **b** -5×-8

   a $4 \times (-5) = -20$ **b** $-5 \times -8 = 40$

You use the same rule for division . . .

- When dividing numbers with different signs, the answer is negative.
- When dividing numbers with the same sign, the answer is positive.

Hint Don't forget that any number that is written without a sign in front of it is always positive.

Example 6

Work out the answers. **a** $-15 \div -5$ **b** $4 \div -2$ **c** $-8 \div 2$

   a $15 \div 5 = 3$ The signs are the same so the answer is positive.

      $-15 \div -5 = 3$

   b $4 \div 2 = 2$ The signs are different so the answer is negative.

      $4 \div -2 = -2$

   c $8 \div 2 = 4$ The signs are different so the answer is negative.

      $-8 \div 2 = -4$

Hint Don't forget that any number that is written without a sign in front of it is always positive.

Exercise 1B

1   Work out the answers.

   a 2×-3 **b** -3×4 **c** -5×2 **d** 6×-3

   e -3×8 **f** -4×5 **g** 3×-4 **h** -6×1

   i 7×-2 **j** 2×8 **k** 6×-10 **l** 8×4

   m -15×2 **n** -6×3 **o** -2×14 **p** -3×11

2   Work out the answers.

   a $12 \div -3$ **b** $-24 \div 4$ **c** $-6 \div 2$ **d** $6 \div -3$

   e $-16 \div 8$ **f** $-20 \div 5$ **g** $12 \div -4$ **h** $-6 \div 1$

   i $10 \div -2$ **j** $12 \div 6$ **k** $60 \div -10$ **l** $8 \div 4$

   m $-14 \div 2$ **n** $16 \div -4$ **o** $-18 \div 3$ **p** $12 \div -6$

3   Find the missing number in each calculation.

   a $2 \times -3 = \ldots$ **b** $-2 \times \ldots = -8$ **c** $3 \times \ldots = -9$

   d $\ldots \times -5 = -15$ **e** $-4 \times 6 = \ldots$ **f** $-3 \times \ldots = -24$

   g $-4 \times \ldots = -20$ **h** $\ldots \times 6 = -36$ **i** $-2 \times 3 = \ldots$

4   Copy and complete each pattern.

   a $3 \times 3 = 9$ **b** $3 \times -2 = -6$ **c** $-3 \times 1 = -3$

      $2 \times 3 = 6$ $2 \times -2 = -4$ $-2 \times 1 = \ldots$

      $1 \times 3 = \ldots$ $1 \times -2 = \ldots$ $-1 \times 1 = \ldots$

      $0 \times 3 = \ldots$ $0 \times -2 = \ldots$ $\ldots \times 1 = \ldots$

      $\ldots \times 3 = \ldots$ $\ldots \times -2 = \ldots$ $\ldots \times 1 = \ldots$

      $\ldots \times 3 = \ldots$ $\ldots \times -2 = \ldots$ $\ldots \times 1 = \ldots$

5   Work out the answers.

   a -2×-4 **b** -3×-5 **c** -5×-5 **d** -6×-4

   e -3×-7 **f** -4×-5 **g** -3×-4 **h** -6×-3

   i -7×-2 **j** -2×-9 **k** -7×-10 **l** -8×-5

   m -12×-2 **n** -6×-5 **o** -3×-11 **p** -4×-12

6 Work out the answers.

    a −9 ÷ −3   **b** −24 ÷ −6   **c** −6 ÷ −2   **d** −16 ÷ −4

    e −12 ÷ −4   **f** −20 ÷ −5   **g** −18 ÷ −6   **h** −5 ÷ −1

    i −10 ÷ −2   **j** −20 ÷ −10   **k** −40 ÷ −10   **l** −8 ÷ −4

    m −14 ÷ −2   **n** −24 ÷ −4   **o** −18 ÷ −9   **p** −30 ÷ −6

7 Find the missing number in each calculation.

    a 12 ÷ … = −3   **b** −24 ÷ … = −8   **c** 21 ÷ … = −7

    d … ÷ −2 = −5   **e** −40 ÷ … = −8   **f** −24 ÷ … = −2

    g −44 ÷ … = 22   **h** … ÷ 6 = −6   **i** … ÷ −3 = 2

8 Work out the answers.

    a 3 × −5   **b** −4 × −6   **c** −6 × 6   **d** −7 × −5

    e 4 × −8   **f** −5 × −6   **g** 4 × −5   **h** −7 × −4

    i 8 × −3   **j** −3 × −10   **k** −8 × −11   **l** −9 × 6

    m −11 × 5   **n** −7 × −6   **o** 4 × −12   **p** −5 × −12

9 Work out the answers.

    a −10 ÷ −2   **b** −24 ÷ 3   **c** −8 ÷ −2   **d** 18 ÷ −3

    e −12 ÷ −6   **f** −20 ÷ 4   **g** −21 ÷ −7   **h** −7 ÷ 1

    i −12 ÷ −3   **j** −25 ÷ −5   **k** 30 ÷ −10   **l** −28 ÷ −4

    m −24 ÷ 2   **n** 44 ÷ −4   **o** −27 ÷ −9   **p** 30 ÷ −5

10 Work out the answers.

    a −8 × −4   **b** −24 ÷ 6   **c** −7 × 5   **d** 11 × −5

    e −15 ÷ −3   **f** −20 × 4   **g** −21 ÷ −3   **h** −8 × 1

    i −11 × −7   **j** −45 ÷ −5   **k** 3 × −12   **l** −80 ÷ −4

    m −14 ÷ 2   **n** 36 ÷ −4   **o** −63 ÷ −9   **p** 12 × −7

Challenge: Algebraic magic squares

Copy and complete each multiplication grid.

A

×	−1	−2	3	4
1				
2				
−3				
−4				

B

×	−1	−3	−5	
1		−3		
2			6	
		−3		
−4				28

C

×				−7
−1		−4		
	6		−15	
−4			20	
		20		

1.3 Factors and highest common factors (HCF)

Learning objective

- To understand and use highest common factors

Key words

common factor

divisible

factor

highest common factor (HCF)

integer

Remember that the **factors** of an **integer** are the integers that divide exactly into it without leaving a remainder (or giving a decimal number as the answer). An integer is a whole number, whether it is positive or negative.

Look at these examples.

- The factors of 6 are 1, 2, 3 and 6. The numbers 4 and 5 are not factors of 6 because when you divide them into 6 you get a remainder (or a decimal answer).
- The factors of 25 are 1, 5 and 25. No other integers divide into 25 exactly.

It is important to remember that every integer (apart from 1) has at least two factors, 1 and itself. For example:

- $1 \times 17 = 17$
- The factors of 17 are 1 and 17.

Another way of saying 'can be divided by' is 'is **divisible** by'. The number 6 is divisible by the integers 1, 2, 3 and 6, and the number 25 is divisible by 1, 5 and 25. Every number is divisible by its factors.

Sets of numbers always have **common factors**. These are numbers that will divide into all of them. For example, the factors of 10 are 1, 2 and 5 and the factors of 15 are 1, 3 and 5. So 5 is a common factor of 10 and 15.

Some pairs of numbers, such as 2 and 3, or 4 and 5, only have 1 as a common factor.

Common factors can help you to solve some mathematical problems, such as simplifying fractions. When numbers have more than one common factor you should use the highest one, which is called the **highest common factor (HCF)**. Using this helps you to simplify a fraction as far as you can.

Example 7

Find the factors of each number.

a 15   **b** 16   **c** 24

  a The factors of 15 are all the numbers that divide into 15 exactly.

    1, 3, 5, 15

  b The factors of 16 are all the numbers that divide into 16 exactly.

    1, 2, 4, 8, 16

  c The factors of 24 are all the numbers that divide into 24 exactly.

    1, 2, 3, 4, 6, 8, 12, 24

Example 8

Find the highest common factor (HCF) of the numbers in each pair.

a 15 and 21   **b** 16 and 24

a Write out the factors of each number.

15: 1, 3, 5, 15

21: 1, 3, 7, 21

You can see that the HCF of 15 and 21 is 3.

b Write out the factors of each number.

16: 1, 2, 4, 8, 16

24: 1, 2, 3, 4, 6, 8, 12, 24

You can see that the HCF of 16 and 24 is 8.

Example 9

Simplify these fractions.   **a** $\frac{18}{30}$ **b** $\frac{15}{27}$

To simplify fractions, you need to divide the numerator (top) and denominator (bottom) by their HCF.

a To simplify $\frac{18}{30}$:   the factors of 18 are 1, 2, 3, 6, 9, 18

the factors of 30 are 1, 2, 3, 5, 6, 10, 15, 30.

The HCF is 6, so divide the numerator and denominator of $\frac{18}{30}$ by 6 to give $\frac{3}{5}$.

b To simplify $\frac{15}{27}$:   the factors of 15 are 1, 3, 5, 15

the factors of 27 are 1, 3, 9, 27.

The HCF is 3, so divide the numerator and denominator of $\frac{15}{27}$ by 3 to give $\frac{5}{9}$.

Exercise 1C

1   Write down all the factors of each number.

a 14	**b** 25	**c** 28	**d** 45	**e** 40
f 50	**g** 35	**h** 36	**i** 60	**j** 100

2   A two-digit number is a factor of 70.

a   Give an example of what the number could be.

b   Now give a different example.

(PS) 3   Sania thinks of a factor of 100. Andrew thinks of a different factor of 100.

The two factors add up to another factor of 100. What could these two factors be?

4   Write down all the common factors for each pair of numbers. The first one has been done for you.

a   10 and 15   **b** 12 and 18   **c** 16 and 20   **d** 25 and 35

> The factors of 10 are 1, 2, 5 and 10.
>
> The factors of 15 are 1, 3, 5, and 15.
>
> So the common factors of 10 and 15 are 1 and 5.

e   30 and 50   **f** 12 and 40   **g** 18 and 24   **h** 27 and 36

(PS) **5** Mrs Bishop is taking 30 pupils on a visit to a museum. She doesn't want to divide them into groups of unequal size. What possible group sizes could she use?

6 Find the HCF of each pair of numbers.

a 12 and 15   b 16 and 24   c 12 and 30   d 18 and 32

e 21 and 28   f 20 and 48   g 18 and 45   h 15 and 22

7 Find the largest number that will divide exactly into both numbers in each pair.

a 30 and 70   b 25 and 35   c 42 and 60   d 14 and 56

e 45 and 60   f 36 and 48   g 50 and 75   h 100 and 160

8 Find the HCF of each pair of numbers.

a 35 and 21   b 16 and 56   c 27 and 54   d 28 and 84

9 Simplify each fraction.

a $\dfrac{10}{15}$ b $\dfrac{10}{30}$ c $\dfrac{16}{24}$ d $\dfrac{18}{27}$ e $\dfrac{12}{64}$ f $\dfrac{21}{28}$ g $\dfrac{45}{75}$ h $\dfrac{36}{54}$

(PS) **10** Miss Speed and Mr Bishop are teachers of two classes in a primary school. They wanted to put the pupils into groups of equal size.

They didn't want to mix the classes.

Miss Speed's class has 32 pupils.

Mr Bishop's class has 24 pupils.

a What is the largest size groups they can make?

b How many of these groups would there be?

Investigation: Tests for divisibility

A How can you tell if a number is divisible by 2?

B **a** Write down some numbers that you know are divisible by 3. Make sure they all have more than one digit.

b Add up the digits of each of these numbers.

c How can you tell if a number is divisible by 3? Write down a rule.

C By looking at the digits of numbers divisible by 4, find a rule for recognising when a number is divisible by 4.

D How can you tell if a number is divisible by 5?

E See how you can combine the rules in parts **A** and **B** to find a rule for recognising when a number is divisible by 6.

F Look at the digits of numbers divisible by 9.

Write down a rule for recognising when a number is divisible by 9.

G Which of these numbers are divisible by:

a 3   b 4   c 6   d 9   e 5?

90, 114, 120, 480, 716, 1503, 111 111

1.4 Multiples and lowest common multiple (LCM)

Learning objective

• To understand and use lowest common multiples

A **multiple** of an integer is the result of multiplying that integer by another integer.

For example, multiplying 3 by 1, 2, 3, 4 and 5 gives 3, 6, 9, 12, 15.

So 3, 6, 9, 12, 15 are all multiples of 3.

This also means that any integer that is divisible by 3, giving another integer without a remainder, must be a multiple of 3.

You can find a **common multiple** for any pair of integers by multiplying one by the other. All pairs of integers will have many common multiples, but the **lowest common multiple (LCM)** is generally the most useful. For example, 3 and 4 both have multiples of 12, 24, 36, ... but the LCM is 12.

Sometimes the LCM is one of the integers. For example, 12 is a multiple of 12 and of 3, but it is also the LCM of 12 and 3.

• $12 = 12 \times 1$

• $12 = 3 \times 4$

You can use LCMs to help in calculations with fractions that have different denominators, as well as in some real-life problems. You will learn more about this in Chapter 12.

Example 10

Find the lowest common multiple (LCM) of each pair of numbers.

a 3 and 7   **b** 6 and 9

  a Write out the first few multiples of each number.

  3: 3, 6, 9, 12, 15, 18, 21, 24, 27, ...

  7: 7, 14, 21, 28, 35, ...

  You can see that the LCM of 3 and 7 is 21.

  b Write out the multiples of each number.

  6: 6, 12, 18, 24, ...

  9: 9, 18, 27, 36, ...

  You can see that the LCM of 6 and 9 is 18.

Example 11

A baker makes small buns, some of mass 15 g and some of mass 20 g.

He wants to sell them in bags that all have the same mass.

What is the smallest mass he could have in each of these bags?

The 15 g cakes could be put into batches of mass 15 g, 30 g, 45 g, 60 g, 75 g, ... (all multiples of 15 g).

The 20 g cakes could be put into batches of mass 20 g, 40 g, 60 g, 80 g, ... (all multiples of 20 g).

The smallest mass will be the lowest common multiple (LCM) of these numbers, which is 60 g.

Exercise 1D

1 Write down the first 10 multiples of:

   a 5   **b** 3   **c** 4   **d** 7   **e** 11.

2 A three-digit number is a multiple of 2.

   a Give an example of what the number could be.

   b Now give a different example.

PS 3 Sean thinks of a multiple of 5.

Gen thinks of a different multiple of 5.

The two numbers added together make a multiple of 10.

What two numbers might they be?

4 Look at the numbers in the box below.

12 15 24 18 45 72 10 55 54 60 100 25

Write down the numbers from the box that are multiples of:

   a 2   **b** 3   **c** 4   **d** 5   **e** 6   **f** 9.

5 Write down the first five multiples of each number.

   a 10   **b** 15   **c** 20   **d** 40   **e** 40

6 **a** Write down the first five multiples of each number.

     i 6   **ii** 8   **iii** 9   **iv** 12

   b Use your answers to part **a** to help find the LCM of the numbers in each pair.

     i 6 and 9   **ii** 6 and 8   **iii** 9 and 12   **iv** 8 and 12

7 Find the LCM of the numbers in each pair.

   a 4 and 6   **b** 6 and 10   **c** 10 and 15   **d** 8 and 14

   e 8 and 20   **f** 9 and 15   **g** 9 and 21   **h** 12 and 16

8 Find the LCM of the numbers in each set.

   a 2, 4 and 5   **b** 3, 4 and 6   **c** 5, 6 and 9   **d** 6, 9 and 10

   e 3, 5 and 20   **f** 5, 6 and 8   **g** 5, 9 and 12   **h** 4, 5, and 6

PS 9 In a year-group of a large school, it is possible to divide the pupils exactly into groups of 28, 30 or 35. Find the smallest number of pupils there could be in this year-group.

PS 10 Two model cars leave the starting line at the same time and travel around tracks of equal lengths. One completes a circuit in 15 seconds, the other in 18 seconds. How long will it be before they are together again at the starting line?

PS **11** Two friends are walking along a straight path.

Arron has a step size of 24 cm and Clinton has a step size of 32 cm.

They both set off, walking from the same point, next to each other. How far will they have gone before they are both in step with each other again?

PS **12** Two sisters regularly go to the same supermarket just after lunch.

Joy goes every six days.

Jess goes every eight days.

How many days in a year are they both likely to be in the supermarket on the same day?

Challenge: LCM and HCF

A **a** Two numbers have an LCM of 20 and an HCF of 4. What are they?

   b Two numbers have an LCM of 18 and an HCF of 3. What are they?

   c Two numbers have an LCM of 60 and an HCF of 5. What are they?

B **a** What are the HCF and LCM of: **i**   5 and 10   **ii**   3 and 12   **iii**   4 and 16?

   b Two numbers, x and y (where y is bigger than x), have an HCF of x.
   What is the LCM of x and y?

C **a** What are the HCF and the LCM of: **i**   5 and 7   **ii**   3 and 4   **iii**   2 and 11?

   b Two numbers, x and y, have an HCF of 1.
   What is the LCM of x and y?

1.5 Squares, cubes and roots

Learning objectives

- To understand and use squares and square roots
- To understand and use cubes and cube roots

Key words	
cube	cube root
square	square root

Squares and roots

You have already met **square** numbers. For example, $5^2 = 5 \times 5 = 25$.

The inverse of squaring a number is finding the **square root**. This is shown by the sign $\sqrt{\ }$. For example, $\sqrt{25} = 5$.

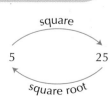

The table gives the square roots of the square numbers up to 225.

You need to learn these.

Square number	1	4	9	16	25	36	49	64	81	100	121	144	169	196	225
Square root	1	2	3	4	5	6	7	8	9	10	11	12	13	14	15

Example 12

Work out the answers.

a $\sqrt{4} \times \sqrt{36}$ **b** $\sqrt{144} \div \sqrt{4}$

a $\sqrt{4} \times \sqrt{36} = 2 \times 6 = 12$

b $\sqrt{144} \div \sqrt{4} = 12 \div 2 = 6$

Cubes and roots

Similarly there are **cube** numbers. You have a cube number when you multiply a number by itself and then multiply by itself again (three 'lots' of the number are multiplied together). For example, $5^3 = 5 \times 5 \times 5 = 125$.

The inverse of cubing a number is finding its **cube root**. It is shown by the sign $\sqrt[3]{\ }$. For example, $\sqrt[3]{125} = 5$.

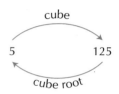

The table gives the cube roots of the cube numbers up to 1000.

It can be useful to recognise these numbers when you see them.

Cube number	1	8	27	64	125	256	343	512	729	1000
Cube root	1	2	3	4	5	6	7	8	9	10

Example 13

Here is a cube with a volume of 64 cm^3.

What is the length of its side?

You know that the volume of the cube is 64 cm^3, and you want to know its side length, so you need to find the **cube root**. This will be the number that, when multiplied by itself and then multiplied by itself again (three 'lots' of the number are multiplied together), gives 64. It is written as $\sqrt[3]{64}$.

$\sqrt[3]{64} = 4$, because $4 \times 4 \times 4 = 64$.

So the side length of the cube is 4 cm.

Note that you can write the square root simply as $\sqrt{\ }$, with no small number 2 in front of it, but the cube root must always have a small 3 in front of it, like this: $\sqrt[3]{\ }$.

Square roots can be positive or negative and a square number is always positive.

A positive cube number can only have a positive cube root and a negative cube number can only have a negative cube root.

$\sqrt[3]{64} = 4$, because $4 \times 4 \times 4 = 64$ but $\sqrt[3]{-64} = -4$, because $-4 \times -4 \times -4 = -64$.

1 Find the value of each number.

  a 3^2 **b** 5^2 **c** $(-2)^2$ **d** 6^2 **e** $(-1)^2$ **f** 4^2

  g 7^2 **h** 8^2 **i** 9^2 **j** 10^2 **k** 12^2 **l** 11^2

2 Use long multiplication to find the value of each number.

  a 13^2 **b** 21^2 **c** 15^2 **d** 17^2 **e** $(-23)^2$ **f** 31^2

  g $(-14)^2$ **h** 18^2 **i** 25^2 **j** 45^2 **k** 19^2 **l** 33^2

3 Find the value of each number. You may have to use long multiplication.

  a 3^3 **b** $(-1)^3$ **c** 5^3 **d** 4^3 **e** 7^3 **f** 8^3

  g 2^3 **h** 9^3 **i** 6^3 **j** 10^3 **k** $(-2)^3$ **l** 11^3

4 Find the area of each square.

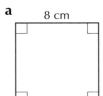

  a 8 cm   **b** 13 cm   **c** 7 cm   **d** 21 cm

5 Find the volume of each cube.

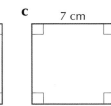

  a 5 cm   **b** 12 cm   **c** 31 cm   **d** 9 cm

6 Work out the answers.

  a $\sqrt{4} \times \sqrt{25}$ **b** $\sqrt{9} \times \sqrt{25}$ **c** $\sqrt{1} \times \sqrt{49}$ **d** $\sqrt{25} \times \sqrt{16}$ **e** $\sqrt{16} \times \sqrt{81}$

  f $\sqrt{64} \times \sqrt{4}$ **g** $\sqrt{100} \times \sqrt{64}$ **h** $\sqrt{36} \times \sqrt{25}$ **i** $\sqrt{49} \times \sqrt{16}$ **j** $\sqrt{81} \times \sqrt{36}$

7 $3^2 + 4^2 = 5^2$ is an example of a special square sum – it is made up of only square numbers. Use your calculator to find which of the following pairs of squares will give you a special square sum.

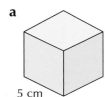

  $5^2 + 12^2$ $3^2 + 7^2$ $6^2 + 8^2$ $7^2 + 12^2$

    $5^2 + 9^2$ $10^2 + 24^2$ $7^2 + 24^2$

8 Work out the answers.

  a $\sqrt{4} \times \sqrt{9}$ **b** $\sqrt{64} \div \sqrt{4}$ **c** $\sqrt{81} \div \sqrt{9}$ **d** $\sqrt{100} \times \sqrt{144}$

  e $\sqrt{25} \times \sqrt{9}$ **f** $\sqrt{49} \times \sqrt{9}$ **g** $\sqrt{25} \times \sqrt{4} \times \sqrt{81}$ **h** $\sqrt{100} \times \sqrt{81} \div \sqrt{36}$

9 Work out the answers.

  a $\sqrt[3]{1} \times \sqrt[3]{8}$ **b** $\sqrt[3]{64} \div \sqrt[3]{8}$ **c** $\sqrt[3]{216} \div \sqrt[3]{27}$ **d** $\sqrt[3]{1000} \times \sqrt[3]{125}$

  e $\sqrt[3]{125} \times \sqrt[3]{729}$ **f** $\sqrt[3]{343} \times \sqrt[3]{27}$ **g** $\sqrt[3]{125} \times \sqrt[3]{512}$ **h** $\sqrt[3]{1000} \times \sqrt[3]{216}$

(PS) 10 What is the volume of a cube with a side area of 9 cm²?

(PS) 11 What is the area of the side of a cube with a volume of 216 cm³?

(MR) 12 Look at what Kim says.

Explain why Kim is correct.

You can't have a square root of a negative number.

(MR) 13 Look at what Lesley says.

Explain why Lesley is not correct.

You can't have a cube root of a negative number.

Challenge: Square roots with a calculator

You can use your calculator to find square roots.

Calculators work in different ways. On some calculators, you have to key the number first and then press the square root key . On others, you have to press the square root key before pressing the number key.

Make sure you know how to use your calculator.

A Use your calculator to find each square root.

    a $\sqrt{12.25}$ **b** $\sqrt{33124}$ **c** $\sqrt{292.41}$ **d** $\sqrt{98.01}$

B Use your calculator to find each square root, to one decimal place.

    a $\sqrt{500}$ **b** $\sqrt{987}$ **c** $\sqrt{75}$ **d** $\sqrt{42}$

1.6 Prime factors

Learning objectives

- To understand what prime numbers are
- To find the prime factors of an integer

Key words	
factor tree	prime factor
prime number	

You may remember that a **prime number** is an integer that has only two factors, itself and one.

These are the first ten prime numbers.

2, 3, 5, 7, 11, 13, 17, 19, 23, 29

A **prime factor** of an integer is a factor that is also a prime number.

Therefore the prime factors of an integer are the prime numbers that will multiply together to give that integer.

Example 14

Find the prime factors of 120.

One way to find the prime factors is to create a **factor tree**, starting with 120 and finding a branch from a prime factor.

Look at the tree.

The first prime factor is 2, so make a branch of 2.

$120 \div 2 = 60$, so you place 60 ready for the next branch.

The next prime factor is 5, so make a branch of 5.

$60 \div 5 = 12$, so you place 12 ready for the next branch.

The next prime factor is 3, so make a branch of 3.

$12 \div 3 = 4$, so you place 4 ready for the next branch.

4 is the product of two primes, 2 and 2, so each have their own branch.

You can now see that the prime factors of 120 are 2, 5, 3, 2 and 2.

Write the factors in increasing order as $2 \times 2 \times 2 \times 3 \times 5$.

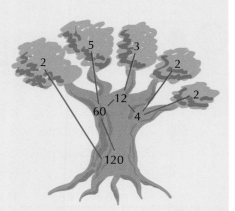

Example 15

Find the prime factors of 18.

Using a prime factor tree, start by splitting 18 into 3×6.

Then split the 6 into 3×2.

You could also split 18 into 2×9 and then split the 9 into 3×3.

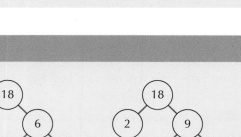

Keep splitting the factors up until you only have prime numbers at the ends of the 'branches'.

Whichever pair of factors you start with, you will always finish with the same set of prime factors.

So, $18 = 2 \times 3 \times 3$.

There is another way to calculate the prime factors of a number.

- Start with the smallest prime number that is a factor of the number.
- Divide that prime number into the integer as many times as possible.
- Then try the next smallest prime number that is a factor of the number.
- Carry on until you reach 1.

Example 16

Find the prime factors of 24.

Use the division method.

```
2 | 24
2 | 12
2 | 6
3 | 3
    1
```

So, $24 = 2 \times 2 \times 2 \times 3$.

Exercise 1F

1 These numbers are written as products of their prime factors. What are the numbers?

   a $2 \times 2 \times 5$ **b** $2 \times 3 \times 5$ **c** $2 \times 2 \times 7$ **d** $3 \times 3 \times 5$ **e** $2 \times 5 \times 5$

   f $3 \times 5 \times 5$ **g** $3 \times 5 \times 7$ **h** $5 \times 5 \times 7$ **i** $3 \times 3 \times 7$ **j** $5 \times 5 \times 5$

2 Use a prime factor tree to work out the prime factors of each number.

   A starting multiplication is given to help you.

   a $42\ (6 \times 7)$ **b** $75\ (5 \times 15)$ **c** $140\ (7 \times 20)$ **d** $250\ (5 \times 50)$

   e $480\ (60 \times 8)$ **f** $72\ (8 \times 9)$ **g** $96\ (4 \times 24)$ **h** $256\ (4 \times 64)$

3 Use a prime factor tree to work out the prime factors of each number.

   a 6   **b** 9   **c** 12   **d** 18   **e** 22

   f 30   **g** 40   **h** 50   **i** 55   **j** 120

4 Use the division method to work out the prime factors of each number.

   a 24   **b** 28   **c** 60   **d** 100   **e** 150

5 Find the prime factors of all the numbers from 2 to 12.

6 **a** Which numbers in Question 5 only have one prime factor?

   b What special name is given to these numbers?

7 10 can be written as a product of its prime factors as $10 = 2 \times 5$.

   a Write down 20 as a product of its prime factors.

   b Write down 50 as a product of its prime factors.

   c Write down 100 as a product of its prime factors.

   d Write down the prime factors of one thousand.

(PS) 8 Find the prime factors of one milllion.

(MR) 9 Alison said: 'The prime factors of 5 are 1 and 5'.

   Explain why Alison is not correct.

Challenge: LCM and HCF diagrams

Make a poster showing the prime factors of a large number, such as 400. Draw a picture of a real tree like the one at the start of this section.

Ready to progress?

I can write down the multiples of any whole number.
I can write down the factors of numbers less than 100.

I can find square and cube numbers and square and cube roots.
I can find common factors for pairs of numbers.

I know that the square roots of positive numbers can have two values, one positive and one negative.
I can find the lowest common multiple (LCM) for pairs of numbers.
I can find the highest common factor (HCF) for pairs of numbers.
I can write a number as the product of its prime factors.

Review questions

1   a   A three-digit number is a multiple of 5.

   i   Give an example of what the number could be.

   ii   Now give a different example.

   b   A two-digit number is a factor of 60.

   i   Give an example of what the number could be.

   ii   Now give a different example.

 2   Matt thinks of a multiple of 6.

Sue thinks of a factor of 30.

The two numbers add together to make 20.

What are the two numbers?

3   Each number below has been written as the product of its prime factors.
What are the numbers?

   a $3 \times 3 \times 5$ b $2 \times 3 \times 5$ c $3 \times 5 \times 5$ d $3 \times 3 \times 11$

4   Look at the list of numbers.

   30 15 5 1 2 6 12 24

   a   Which of the numbers are multiples of:
      i   3   ii   5   iii   30?
   b   Which of the numbers are factors of:
      i   10   ii   30   iii   36?

5   Work out the answers.

    a -7×-5 **b** $-25 \div 5$ **c** -8×4 **d** 12×-4

    e $-16 \div -2$ **f** -21×3 **g** $-21 \div -7$ **h** -9×1

    i -12×-6 **j** $-55 \div -11$ **k** 4×-11 **l** $-60 \div -3$

    m $-18 \div 3$ **n** 16×-4 **o** $-54 \div -6$ **p** 12×-8

 6   Explain why:

    a   not all odd numbers are prime numbers.

    b   all even numbers are multiples of two.

    c   all prime numbers have exactly two factors.

    d   all prime numbers between 20 and 30 are odd numbers.

7   Put these values in order of size, with the smallest first.

    7^2 2^3 $\sqrt{25}$ $\sqrt[3]{27}$

8   Work out the areas of these squares.

a   6 cm   **b**   32 cm

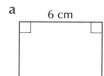

9   Work out the volumes of these cubes.

a   **b**

4 cm   14 cm

 10   Hannah is taking 60 students on a trip to Rome. She doesn't want to divide them into groups of unequal size, but wants them in groups larger than 4. What possible group sizes could she use?

 11   Find the LCM of each pair of numbers.

    a   5 and 7   **b**   6 and 18   **c**   24 and 36

Challenge
The Eiffel Tower

The Eiffel Tower was built in Paris for the 1889 World's Fair. It was built to mark the 100-year anniversary of the French Revolution.

Over 18 000 pieces of puddle iron and 2.5 million rivets were used to make the tower. Puddle iron is a type of wrought iron used in construction. 300 workers spent two years fixing together the framework of the tower, which stood nearly 1000 feet high and was the tallest structure in the world. It was the tallest structure until New York City's Chrysler Building was built in 1930. In 1957, an aerial was added that increased the structure's height by 65 feet. This made it taller than the Chrysler Building but not the Empire State Building, which was built in 1931.

The Eiffel Tower had a major facelift in 1986 and is now repainted every seven years.

It now has more visitors than any other paid monument in the world – about 7 million people per year.

There are 500 employees who work there, working in its restaurants, operating its lifts, making sure of its security and directing the crowds.

1 When was the French Revolution?

2 When did the Eiffel Tower celebrate its 100th anniversary?

3 When will it celebrate its 150th anniversary?

4 How many times has the tower been painted since 1986?

5 The Eiffel Tower is open every day of the year. What is the average number of people visiting the tower each day?

6 A rivet has an average mass of 15 g. What is the total mass of all the rivets in the Eiffel Tower?

7 Each piece of puddle iron has an average mass of 75 kg. What is the total mass of the puddle iron in the framework of the Eiffel Tower. Give your answer in tonnes.

8 A foot is approximately equivalent to 30.5 cm. Calculate how tall the Eiffel Tower is now in metres.

9 In 2014, it cost €14 to go to the Eiffel Tower and £30 to visit Blackpool Tower.
The exchange rate in 2014 was £1 = €1.20.
Which tower was the cheapest to visit and by how much? Give your answer in pounds and pence.

2

Geometry

This chapter is going to show you:

- how to identify parallel and perpendicular lines
- how to work out angles in triangles and quadrilaterals
- how to translate a shape
- how to rotate a shape.

You should already know:

- the names of the different types of triangle and quadrilateral
- that the angles on a straight line add up to 180°
- how to plot coordinates.

About this chapter

Architects use the properties of angles and shapes to design buildings. When you look at a building you can often see triangles, quadrilaterals and other shapes.

Understanding angles, lines and shapes and being able to construct them accurately are important in all types of design. For example, many fabrics and wallpapers use angles, lines and shapes that have been rotated, reflected and repeated to make a pattern that looks good and is also easy to reproduce. When fabrics and wallpaper are printed, the design is reproduced by translating it repeatedly.

This chapter will show you some of the properties of angles and shapes.

2.1 Parallel and perpendicular lines

Learning objectives

- To identify parallel lines
- To identify perpendicular lines

These two lines are **parallel**.

When you extend the lines in both directions, they never meet.

You can show that two lines are parallel by drawing arrows on them like this:

Two lines are **perpendicular** when the angle between them is 90°.

90°

The angle formed where lines that are perpendicular to each other meet called a **right angle**. You can show that two lines are perpendicular by labelling the 90° angle with a square corner.

Exercise 2A

1 Write down which sets of lines are parallel.

a

b

c

d

e

f

g

h

2 Copy each diagram onto square dotted paper.

On each diagram, use your ruler to draw two more lines that are parallel to the first line.

Show that the lines are parallel by adding arrows to them.

a

b

c

d

e

f

3 Write down which pairs of lines are perpendicular.

a

b

c

d

e

f

g

h

4 The line AB is perpendicular to the line CD.

Copy and complete each sentence.

a

X   Y

Z

The line XY is perpendicular to the line...

b

R   S

P

Q

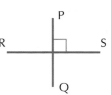

The line PQ is perpendicular to the line...

c

A   B

C   D

The line CD is perpendicular to the lines... and...

d

E   F

H   G

The line EF is perpendicular to the lines... and...

5 Copy each diagram onto square dotted paper. Add arrows and square corners to show which lines are perpendicular and which are parallel to each other.

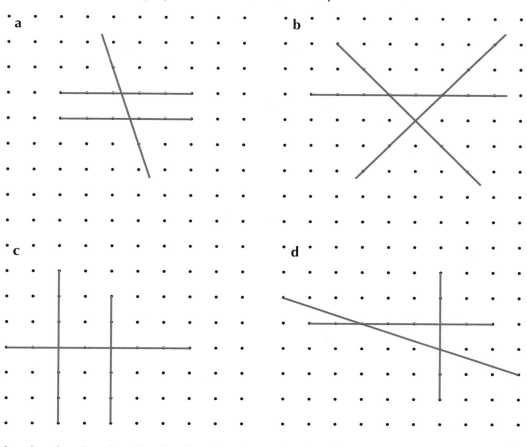

a

b

c

d

6 For the shape ABCDE:

a write down the sides that are equal in length

b write down the angles that are equal in size

c write down the sides that are parallel

d write down the sides that are perpendicular to each other.

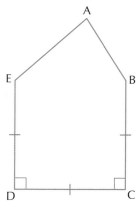

2.2 Angles in triangles and quadrilaterals

Learning objectives

- To know that the sum of the angles in a triangle is 180°
- To know that the sum of the angles in a quadrilateral is 360°

Angles in a triangle

You can draw and cut out a triangle ABC from coloured plain paper.

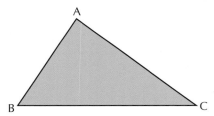

Then tear off the three angles at A, B and C.

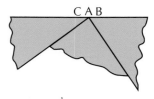

Now place the angles together.

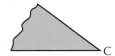

What do you notice?

You should see that the three angles make a straight line.

You already know that the angles on a straight line add up to 180°.

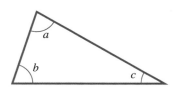

This shows that the angles in any triangle add up to 180°.

$a + b + c = 180°$

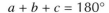

Example 1

a Work out the size of the angle labelled *a* in this scalene triangle.

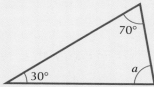

b Work out the size of the angle labelled *b* in this right-angled triangle.

c Work out the size of the angle labelled *c* in this isosceles triangle.

a $a = 180° - 70° - 30°$

$a = 80°$

b $b = 180° - 90° - 50°$

$b = 40°$

c The triangle is an isosceles triangle so the other angle is also 70°.

So $c = 180° - 70° - 70°$.

$c = 40°$

Angles in a quadrilateral

Cut out a quadrilateral ABCD.

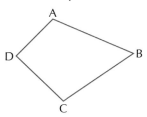

Cut the quadrilateral into two triangles.

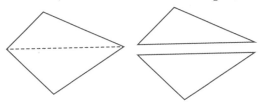

The three angles in each triangle add up to 180°.

The six angles in the two triangles add up to 360° and these make up the four angles in the quadrilateral.

So the angles in any quadrilateral add up to 360°.

In the diagram, $a + b + c + d = 360°$.

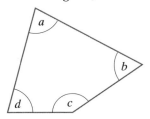

Example 2

Work out the size of the angle labelled a.

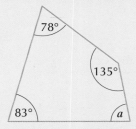

The angles in a quadrilateral add up to 360°.

So $a = 360° - 135° - 78° - 83°$.

$a = 64°$

Exercise 2B

1. Work out the size of each unknown angle in these scalene triangles.

   a   **b**   **c**   **d**

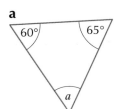
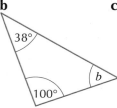
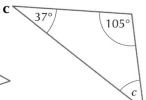
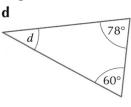

2. Work out the size of each unknown angle in these right-angled triangles.

   a   **b**   **c**   **d**

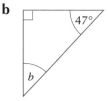
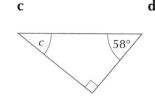
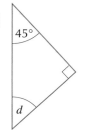

3 Work out the size of each unknown angle in these isosceles triangles.

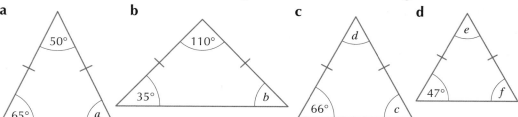

a

50°

65°   a

b

110°

35°   b

c

d

66°   c

d

e

47°   f

4 Work out the size of each unknown angle.

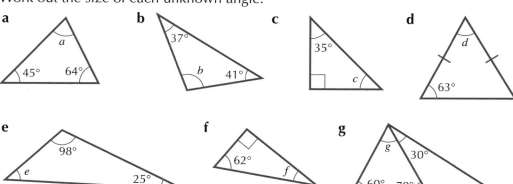

a

a

45° 64°

b

37°

b   41°

c

35°

c

d

d

63°

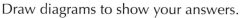

e

98°

e   25°

f

62°

f

g

g   30°

60° 70°

h

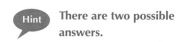

(MR) **5** One angle in an isosceles triangle is 70°.

Work out the possible sizes of the other two angles.

Draw diagrams to show your answers.

Hint There are two possible answers.

(PS) **6** In this triangle, the size of the angle labelled *b* is 20° more than the size of the angle labelled *a*.

The value *c* is 20° more than the value of *b*.

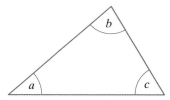

b

a   c

Work out the size of each angle.

7 Work out the size of each unknown angle.

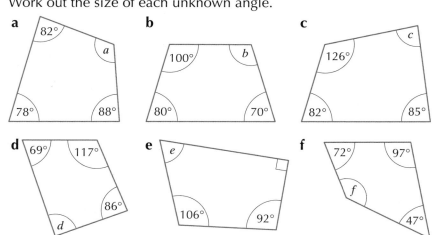

a

82°

a

78°   88°

b

100°   b

80°   70°

c

c

126°

82°   85°

d

69°   117°

86°

d

e

e

106°   92°

f

72°   97°

f

47°

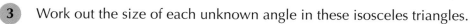

8 Work out the size of each unknown angle.

a

b
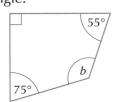

c
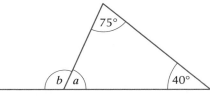

d
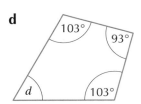

Reasoning: Interior and exterior angles of triangles

In the diagram the angle labelled *a* is called an interior angle and the angle labelled *b* is called an exterior angle.

Work out the size of the angles marked *a* and *b*.

The reasons are given.

$a = 180° − 75° − 40°$

So $a = 65°$.   Angles in a triangle add up to 180°.

$b = 180° − 65°$

So $b = 115°$.   Angles on a line add up to 180°.

In each diagram below, state which angle is the interior angle and which angle is the exterior angle.

Work out the size of each unknown angle.

A
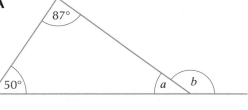

B
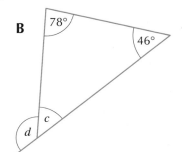

C
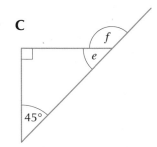

2.3 Translations

Learning objective

- To understand how to translate a point or a shape

Key words

translate

translation

A **translation** is the movement of a point or the movement of a 2D shape from one position to another.

To **translate** a point or shape you move the stated number of unit squares to the *right or left*, followed by the stated number of unit squares *up or down*.

The distance and direction of the translation is described by the number of unit squares moved to the right or left, followed by the number of unit squares moved up or down.

As with reflections and rotations, the original shape is the object and the translated shape is the image.

Example 3

Write down the translation that moves:

a point A onto point B　　**b** point C onto point D　　**c** point E onto point F.

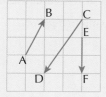

   a Point A moves onto point B by the translation 1 unit right and 2 units up.

   b Point C moves onto point D by the translation 2 units left and 3 units down.

   c Point E moves onto point F by the translation 2 units down.

Example 4

Translate triangle A onto triangle B by the translation 3 units right and 2 units up.

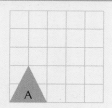

Points on triangle A are translated onto triangle B, as shown by the arrows.

When an object is translated onto its image, every point on the object moves the same distance.

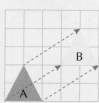

Example 5

Describe the translation of rectangle ABCD.

   The rectangle ABCD has translated onto rectangle A′B′C′D′ by the translation 3 units left and 3 units down.

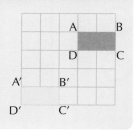

1. Write down the translation that moves:

   a point A onto point B

   b point C onto point D

   c point E onto point F

   d point G onto point H.

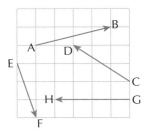

2. On a copy of the grid, show the translation that moves point A:

   a 3 units right and 2 units up to a point B

   b 2 units right and 4 units down to a point C

   c 1 unit left and 2 units up to a point D

   d 3 units left and 3 units down to a point E.

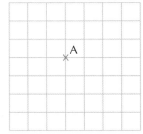

3. Describe the translation:

   a from A to B   b from B to C   c from A to E

   d from E to C   e from E to D.

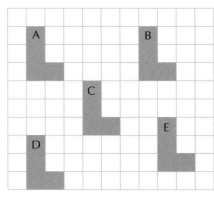

4. Copy the triangle ABC onto a square grid. Label it X.

   a Write down the coordinates of the vertices of triangle X.

   b Translate triangle X 2 units right and 4 units up. Label the new triangle Y.

   c Write down the coordinates of the vertices of triangle Y.

   d Translate triangle Y 3 units right and 3 units down. Label the new triangle Z.

   e Write down the coordinates of the vertices of triangle Z.

   f Describe the translation that translates triangle Z onto triangle X.

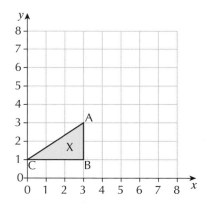

(MR) 5. Shape P moves to shape Q by the translation 4 units left and 5 units up.

   What translation moves shape Q to shape P?

(PS) 6. Without drawing a coordinate grid, write down the translation that moves:

   a the point A(1, 2) to the point B(3, 6)

   b the point C(3, 5) to the point D(6, 3)

   c the point E(5, 8) to the point F(2, 2).

Investigation: Dotty translations

A right-angled triangle is drawn on a 3 by 3 grid of dots.

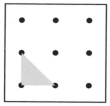

The triangle is to be translated from dot to dot.
Here is an example.

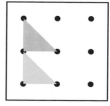

A How many different translations of the triangle are possible on the grid?

B How many different translations of the triangle are possible on this 4 by 4 grid of dots?

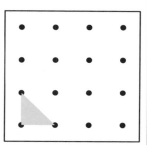

2.4 Rotations

Learning objective

- To understand how to rotate a shape

Key words

| angle of rotation |
| centre of rotation |
| direction of rotation |
| rotation |

You have seen how a 2D shape can be rotated.

To describe the **rotation** of a 2D shape accurately, you need to know three facts. These are:

- the **centre of rotation** – the point about which the shape rotates
- the **angle of rotation** – usually 90° (a quarter-turn) or 180° (a half-turn)
- the **direction of rotation** – clockwise (to the right) or anticlockwise (to the left).

When you rotate a shape, it is helpful to use tracing paper.

As with reflections, the original shape is called the object and the rotated shape is called the image.

Example 6

Describe the rotation of this flag.

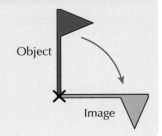

The flag has been rotated through a quarter-turn clockwise about the point X.

You can also say that the flag has been rotated through 90° clockwise about the point X.

Example 7

Describe the rotation of this triangle.

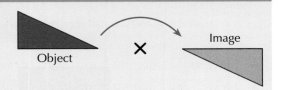

The right-angled triangle has been rotated through a half-turn clockwise about the point X.

You can also say that the right-angled triangle has been rotated through 180° clockwise about the point X.

Notice that the object can be rotated either clockwise or anticlockwise about the centre of rotation, when turning through 180° to form the image.

Example 8

Describe the rotation of triangle ABC.

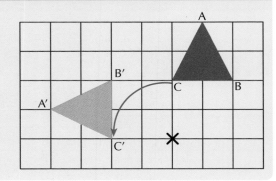

Triangle ABC has been rotated through 90° anticlockwise, about the point X, onto triangle A'B'C'.

Notice that the rotated triangle is labelled in the same way as with reflections.

Example 9

a Rotate triangle ABC through 90° clockwise about the point X on the grid.

Label the image triangle A'B'C'.

b Write down the coordinates of the vertices of the object triangle and the image triangle.

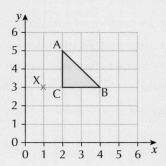

a Triangle A′B′C′ is the image of triangle ABC after a rotation of 90° clockwise about the point X.

A and A′ are the same distance from X.

B and B′ are the same distance from X.

C and C′ are the same distance from X.

The directions of A and A′ from X are at 90° to each other, which is the angle of rotation.

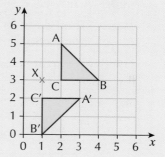

b The coordinates of the vertices of the object are A(2, 5), B(4, 3) and C(2, 3).

The coordinates of the vertices of the image are A′(3, 2), B′(1, 0) and C′(1, 2).

Exercise 2D

1 Write down the angle of rotation and the direction for each rotation.

a 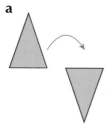   **b** 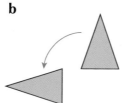   **c**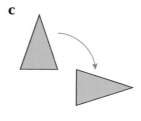

2 Copy each flag below and draw its image after it has been rotated about the point marked X, through the angle indicated. Use tracing paper to help.

a    **b**    **c**

    a quarter-turn   a quarter-turn   a half-turn
     clockwise   anticlockwise   clockwise

3 Copy each arrow below and draw its image after it has been rotated about the point marked X, through the angle indicated. Use tracing paper to help.

a   **b**   **c**

   90° clockwise   90° anticlockwise   180° clockwise

4 Copy each shape onto a square grid.

Then draw its image after it has been rotated, about the point marked X, through the angle shown. Use tracing paper to help.

a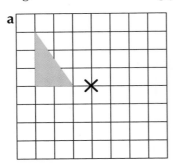

b

c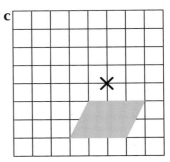

90° clockwise   90° anticlockwise   180° anticlockwise

 5 Copy triangle A onto a square grid.

a Rotate triangle A through 90° clockwise about the point X. Label the triangle B.

b Rotate triangle B through 90° clockwise about the point X. Label the triangle C.

c Rotate triangle C through 90° clockwise about the point X. Label the triangle D.

d Describe the rotation that moves triangle D onto triangle A.

e Describe the rotation that moves triangle D onto triangle B.

f Describe the rotation that moves triangle C onto triangle B.

6 Copy triangle A onto a coordinate grid.

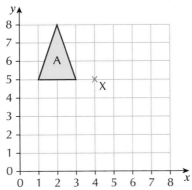

a Rotate triangle A through 180° clockwise about the point X(4, 5).
Label the new triangle B.

b Now translate triangle B 4 units left and 2 units down.
Label the new triangle C.

c Describe the rotation that moves triangle C onto triangle A.

7 Copy triangle PQR onto a coordinate grid.

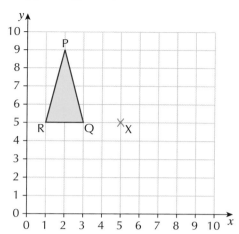

a Write down the coordinates of the vertices of triangle PQR.

b Rotate triangle PQR through 90° clockwise about the point X. Label it triangle P′Q′R′.

c Write down the coordinates of the vertices of triangle P′Q′R′.

d Rotate triangle PQR through 90° anticlockwise about the point X. Label it triangle P″Q″R″.

e Write down the coordinates of the vertices of triangle P″Q″R″.

Challenge: Finding the centre of rotation

The coordinate grid shows three identical isosceles triangles A, B and C.

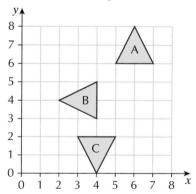

A Triangle A is rotated onto triangle B through 90° anticlockwise.

> **Hint** The centre of rotation is different for both triangles.

Use tracing paper to find the coordinates of the centre of rotation.

B Triangle A is rotated onto triangle C through 180° anticlockwise.

Use tracing paper to find the coordinates of the centre of rotation.

Ready to progress?

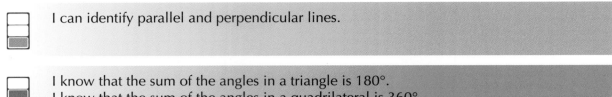

I can identify parallel and perpendicular lines.

I know that the sum of the angles in a triangle is 180°.
I know that the sum of the angles in a quadrilateral is 360°.
I can translate a point or a 2D shape.
I can rotate a 2D shape.

Review questions

1 Copy each shape onto squared paper. Add square corners and arrows to show which lines are perpendicular and which are parallel to each other.

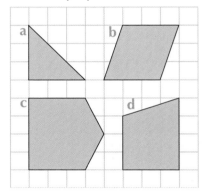

2 Work out the size of each unknown angle.

a

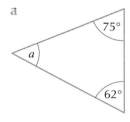

b

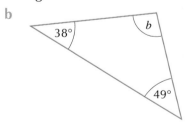

c

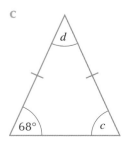

d

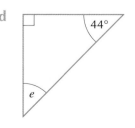

3 Work out the size of each unknown angle.

a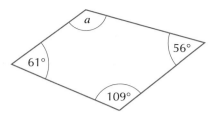

b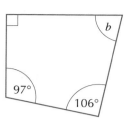

(MR) **4** State whether the following statements are true or false.

a   A quadrilateral can have one right angle.

b   A quadrilateral can have two right angles.

c   A quadrilateral can have three right angles.

d   A quadrilateral can have four right angles.

5 Copy the trapezium ABCD onto a coordinate grid.

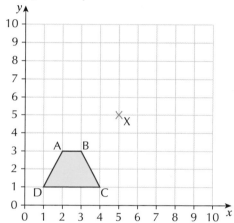

a   Write down the coordinates of the vertices of the trapezium ABCD.

b   Rotate the trapezium ABCD through 180° clockwise about the point X.
Label it A'B'C'D'.

c   Write down the coordinates of the vertices of the trapezium A'B'C'D'

d   Describe the rotation that moves the trapezium A'B'C'D' onto the trapezium ABCD.

6 Describe the translation:

a   from A to B

b   from C to D

c   from B to C

d   from C to A.

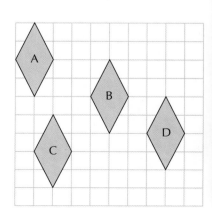

Challenge
Constructing triangles

You need to be able to use a ruler, a protractor and compasses to draw shapes exactly, from the information you are given. This is called constructing a shape.

When constructing a shape you must draw lines accurately to the nearest millimetre and angles to the nearest degree.

Example: How to construct a triangle if you know two angles and one side.

Construct the triangle XYZ.

* Draw a line YZ 8 cm long.

* Draw an angle of 40° at Y.

* Draw an angle of 50° at Z.

* Extend both angle lines to intersect at X to complete the triangle.

The completed, full-sized triangle is given here.

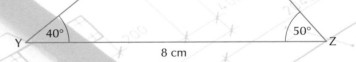

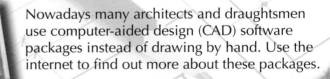

Nowadays many architects and draughtsmen use computer-aided design (CAD) software packages instead of drawing by hand. Use the internet to find out more about these packages.

Example: How to construct a triangle if you know two sides and one angle.

Construct the triangle ABC.

- Draw a line BC 7 cm long.
- Draw an angle of 50° at B.
- Draw the line AB 4 cm long.
- Join AC to complete the triangle.

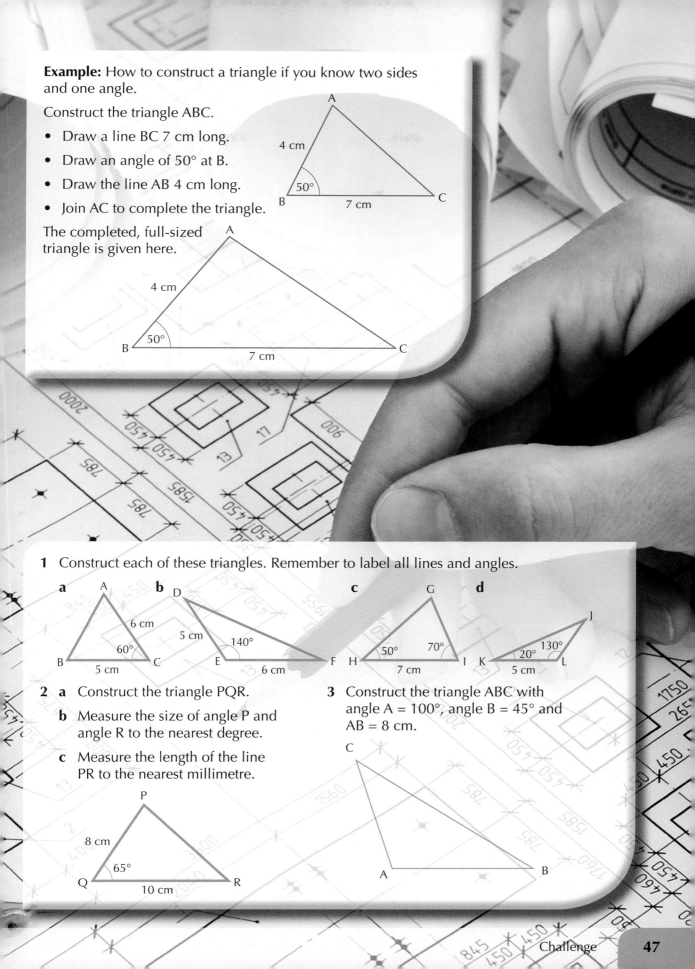

The completed, full-sized triangle is given here.

1 Construct each of these triangles. Remember to label all lines and angles.

a **b** **c** **d**

2 a Construct the triangle PQR.

b Measure the size of angle P and angle R to the nearest degree.

c Measure the length of the line PR to the nearest millimetre.

3 Construct the triangle ABC with angle A = 100°, angle B = 45° and AB = 8 cm.

3

Probability

This chapter is going to show you:

- how to work with a probability scale
- how to work with the probability of an event not happening
- how to work out probabilities using sample spaces where necessary
- how to collect data from a simple experiment and record it in a frequency table.

What you should already know:

- what chance and probability are
- how to collect data from a simple experiment
- how to record data in a table or chart.

About this chapter

What is the probability that you will ever travel in space?

One hundred years ago, the chance of this was nil, that is, it was impossible, but now the chance is increasing every decade. Scientists predict that many pupils in schools now will have a fair chance of travelling into space one day in their lifetime. They calculate the probabilities by working out what is technically possible, and who might be able to afford it.

We do not know for certain if mass space travel will happen but, by studying probability, we can understand how likely it is to happen and how the scientists work it out.

3.1 Probability scales

Learning objective

- To use a probability scale to represent a chance

Key words

equally likely	event
outcome	probability
probability scale	

When you do something such as rolling a dice, this is called an **event**. The possible results of the event are called its **outcomes**. For example, rolling a dice has six possible outcomes: scoring 1, 2, 3, 4, 5 or 6.

The words used to describe how likely an event is to happen are:

impossible, very unlikely, unlikely, even chance, likely, very likely, certain

You can use **probability** to decide how likely it is that different outcomes will happen.

Equally likely outcomes

Equally likely outcomes are those that all have the same chance of happening. For example, when you roll a dice, there are six different possible outcomes because it could land so that any one of its six numbers shows on top. When it is a fair dice the chances of the outcome being any one of the six numbers are equal.

The probability of an equally likely outcome is:

$$P(\text{outcome}) = \frac{\text{the number of ways the outcome could occur}}{\text{the total number of possible outcomes}}$$

There is only one way a normal dice can show a 6 when it lands, so only one of its possible outcomes will be the one you want.

There are six numbers on the faces, so there are six possible outcomes. Therefore:

$$P(6) = \frac{1}{6}$$

Example 1

There are 10 coloured counters in a bag. Three are blue, five are red and the rest are yellow. A counter is picked out at random. Calculate the probability of the counter being:

a blue   **b** red   **c** yellow.

a As there are 3 blue counters out of 10 counters altogether, the probability of getting a blue counter is $\frac{3}{10}$.

b As there are 5 red counters out of 10 counters altogether, the probability of getting a red counter is $\frac{5}{10} = \frac{1}{2}$.

c There are $(10 - 3 - 2) = 2$ yellow counters in the bag.

As there are 2 yellow counters out of 10 counters altogether, the probability of getting a yellow counter is $\frac{2}{10} = \frac{1}{5}$.

Probabilities can be written as either fractions or decimals. They always take values between 0 and 1, including 0 and 1. The probability of an event happening can be shown on a **probability scale**.

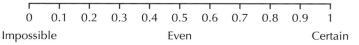

0	0.1	0.2	0.3	0.4	0.5	0.6	0.7	0.8	0.9	1

Impossible   Even   Certain

Example 2

Terry has a bag of marbles.

In the bag are one red marble, two green marbles, three white marbles and four blue marbles.

Terry took one marble out of the bag at random. Show on a probability scale the probability that it is:

a white   **b** green   **c** red   **d** blue.

There are $1 + 2 + 3 + 4 = 10$ marbles in the bag.

a 3 marbles are white, so the probability of taking out a white marble is $\frac{3}{10} = 0.3$.

b 2 marbles are green, so the probability of taking out a green marble is $\frac{2}{10} = 0.2$.

c 1 marble is red, so the probability of taking out a red marble is $\frac{1}{10} = 0.1$.

d 4 marbles are blue, so the probability of taking out a blue marble is $\frac{4}{10} = 0.4$.

Show the probabilities on a probability scale.

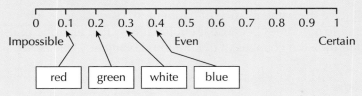

Probabilities of events not occurring

Sam has a bag of ten counters. Seven of the counters are blue.

He picks a counter out of the bag.

The probability that he picks a blue counter is:

P(picking a blue counter) $= \frac{7}{10} = 0.7$.

The probability that he does not pick a blue counter is:

P(not picking a blue counter) $= \frac{3}{10} = 0.3$.

When one outcome is the absolute opposite of another outcome, such as 'picking a blue counter' and 'not picking a blue counter', then the probabilities of the two outcomes add up to 1.

So you could also find P(not picking a blue counter) by:

$1 - $ (picking a blue counter) $= 1 - 0.7 = 0.3$.

P(event NOT happening) $= 1 - $ P(event happening)

Example 3

The probability of it raining in Manchester during July is 0.8.

What is the probability of it NOT raining in Manchester during July?

The probability of it not raining, or P(not raining), is:

$1 - 0.8 = 0.2$.

So the chance of it not raining in Manchester during July is 0.2.

Exercise 3A

1 Josh has six blue shirts, three white shirts and one grey shirt in his wardrobe. He picks a shirt at random.

Copy and complete each sentence.

a It is likely that Josh will pick a ...... shirt.

b It is very unlikely that Josh will pick a ...... shirt.

c It is ...... that Josh will pick a green shirt.

2 Dave has some coins in his pocket. He has three one penny coins, six two pence coins and one ten pence coin.

He drops a coin when he takes out his handkerchief.

Copy and complete each sentence.

a It is likely that Dave has dropped a ...... coin.

b It is very unlikely that Dave has dropped a ...... coin.

c It is ...... that Dave will drop a one penny coin.

3 Ten cards are numbered 0 to 9.

A card is picked at random. Work out the probability that it is:

a 2   **b** not 2   **c** odd   **d** not odd

There are 10 cards altogether. There is one card with a 2 on it.

The probability is $\frac{1}{10}$.

e 7, 8 or 9   **f** less than 7   **g** 4 or 5   **h** not 4 or 5?

The first one is done for you.

4 Toni has a bag of jelly babies. Seven are red, two are green and one is black.

a What colour jelly baby is Toni most likely to pick out?

b What is the probability of Toni picking at random:

     i a black jelly baby   **ii** a green jelly baby   **iii** a red jelly baby?

5 Mr Bradshaw has a box of 20 calculators. Three of these do not work.

a How many calculators in Mr Bradshaw's box do work?

b What is the probability of taking a calculator out, at random, and it:

     i working   **ii** not working?

6 The coastal town of Redpool has ten small trains that run along the sea front. There are two red trains, five blue trains and three green trains.

Ali is waiting for a train to arrive. Calculate the probability that the train is:

a green   **b** red   **c** red or blue   **d** yellow

e not green   **f** not red   **g** neither red nor blue   **h** not yellow.

7 The cellar of a café was flooded in the floods in early 2014. All the labels came off the tins of soup and were floating in the water.

The café owner knew that she had 17 cans of mushroom soup, 15 cans of tomato soup, 10 cans of vegetable soup and 8 cans of pea soup.

a How many tins of soup were there in the flooded cellar?

b After the flood, what is the probability that a tin of soup chosen at random is:

    i mushroom   **ii** tomato   **iii** vegetable   **iv** pea?

8 Here is a probability scale.

The probability of events A, B, C and D happening are shown on the scale. Copy the scale and mark underneath it the probabilities of A, B, C and D *not* happening.

9 In a town there are 24 red buses, 6 blue buses and 10 green buses. Calculate the probability that the next bus to arrive at a bus stop is:

a green   **b** red   **c** red or blue   **d** yellow

> There are 24 + 6 + 10 = 40 buses altogether. There are 10 green buses.
>
> The probability is $\frac{10}{40} = \frac{1}{4}$.

The first one is done for you.

e not green   **f** not red   **g** neither red nor blue   **h** not yellow.

10 For a school project, Joy asks her friends to choose a random number between 1 and 40.

What is the chance that a number given to her:

a is odd   **b** will only divide by itself exactly   **c** has a 5 in it

d has a 0 in it   **e** will divide exactly by 3   **f** is a square number

g is bigger than 10   **h** will divide exactly into 9   **i** will divide exactly into 40.

(MR) **11** The chance of it snowing in February is 0.6. Explain why the probability of it not snowing in February is 0.4.

Activity: Choosing numbers

A Look at the numbers in this table. A number is chosen at random. What is the probability that it is:

a even   **b** a multiple of 5   **c** a factor of 36?

3	4	7	9
10	13	15	16
18	20	21	26
30	35	36	38
41	45	46	50

B Make up your own table of numbers and your own set of questions.

Then answer them or swap them with classmates.

3.2 Collecting data for a frequency table

Learning objectives

- To collect data and use it to find probabilities
- To decide if an event is fair or biased

Handling data is about collecting and organising information. It is also about presenting data using diagrams and being able to interpret diagrams.

Example 4

A spinner has five different coloured sections on it: red, green, blue, yellow and black. Alex wants to test whether the spinner is **fair**.

Alex draws up a table like the one shown. She records which colour the spinner lands on each time she spins it. Her table shows the outcome each time the spinner is spun. She records each outcome with a tally mark.

Colour	Tally	Total
Red	III	3
Green		0
Blue	II	2
Yellow	II	2
Black	III	3

One conclusion from these results is that the spinner might be **biased** (not fair) because it never landed on green.

To be more certain, she would need to carry out many more trials – 50 spins at least!

Example 5

Use the results in Example 4 to estimate the probability of the spinner:

a landing on red   **b** landing on green   **c** landing on blue.

  a The spinner lands on red 3 times out of 10, so the estimate of P(red) is $\frac{3}{10}$.

  b The spinner lands on green 0 times out of 10 so the estimate of P(green) is 0.

  c The spinner lands on blue 2 times out of 10, so the estimate of P(blue) = $\frac{2}{10}$ = 0.2.

Exercise 3B

  1 The table shows the results of spinning a spinner with three coloured sections.

Colour	Tally	Total
Blue	II	
Yellow	III	
Black	HHt	

  a Use the results to estimate the probability of the spinner landing on:

    i blue   **ii** yellow   **iii** black.

  b Comment on whether the spinner is fair or not.

MR **2** The table shows the results of spinning a spinner with four numbered sections.

Number	Tally	Total
0	II	
1	IIII	
2	III	
3	I	

a Use the results to estimate the probability of the spinner landing on:

i 0   **ii** 1   **iii** 2   **iv** 3   **v** 4.

b Comment on whether you think the spinner is biased or not.

MR **3** The table shows the results of rolling a 6-sided dice.

Number	Tally	Total
1	III	
2	IIII	
3	III	
4	III	
5	III	
6	IIII	

a Use the results to estimate the probability of rolling a:

i 1   **ii** 2   **iii** 3   **iv** 4   **v** 5   **vi** 6   **vii** 7.

b Comment on whether you think the dice is biased or not.

MR **4** Make your own spinner from a piece of card. It can have five sections, as in Example 4, or a different number of sections. Label the sections with numbers or colour each in a different colour.

Spin the spinner 50 times and record the results in a table. Comment on whether you think that your spinner is biased.

5 Put 20 counters of different colours in a bag.

a Draw out a counter, note the colour and replace it. Repeat this 50 times. Record your results in a tally chart.

b Use your results to estimate the probability of choosing each colour. Check your results by emptying the bag.

MR **6** **a** Roll a dice 60 times. Record your results in a table.

b Use the results to estimate the probability of each score.

c Comment on whether you think the dice is biased.

7 **a** Flip a coin 40 times. Record the results in a table.

b Use the results to check that the probability of flipping a coin and getting a head is 0.5.

Activity: Rolling two dice

When you roll two dice, you have exactly the same chance of rolling a total greater than 8 than you have of rolling a total less than 6.

Roll two dice sufficient times, keeping a record of the results, to see if this is true.

3.3 Mixed events

Learning objective

• To recognise mixed events where you can distinguish different probabilities

If you have a mixed event, when there are several possible choices, you must be careful to look at exactly how many possible outcomes there are. You must also consider how many ways any particular outcome can happen. The next example shows you how to do this.

Example 6

Mrs Smith can choose a banana, a pear or an apple.

What is the probability of her choosing:

a a banana   **b** a green apple   **c** an apple   **d** a green fruit?

First count the number of each fruit.

There are 20 green apples, 25 red apples, 15 green pears and 40 bananas.

So there are 20 + 25 + 15 + 40 = 100 pieces of fruit altogether.

There are also 45 apples and 35 pieces of green fruit.

a   P(banana) = $\frac{40}{100}$ = 0.4

b   P(green apple) = $\frac{20}{100}$ = 0.2

c   P(apple) = $\frac{45}{100}$ = 0.45

d   P(green fruit) = $\frac{35}{100}$ = 0.35

Exercise 3C

Use the 20 yachts in the picture to answer the questions 1 to 3 below.

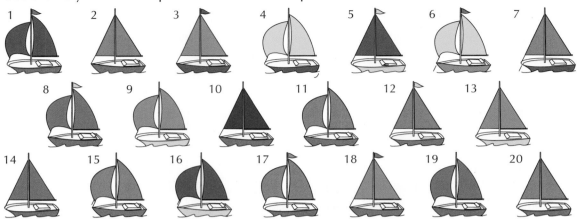

1 How many yachts have:
    a round sails   **b** a blue hull   **c** yellow sails
    d a flag at the top of the mast **e** a pointed hull   **f** red sails and a curved hull
    g a yellow hull but no flag   **h** straight green sails **i** a curved red hull
    j a red hull, blue sails and a flag at the top of the mast?

2 A sailor takes a yacht at random. What is the probability that it has:
    a round sails   **b** a blue hull   **c** yellow sails
    d a flag at the top of the mast **e** a pointed hull   **f** red sails and a curved hull
    g a yellow hull but no flag   **h** straight green sails **i** a curved red hull
    j a red hull, blue sails and a flag at the top of the mast?

3 All 20 boats were in a race. Three friends were on the boats.

Peter was on a boat with green sails.

Ali was on a boat with a blue hull.

Andy was on a boat with straight sails.

    a A boat with a blue hull won the race.

       What was the probability that Ali was on the boat that won the race?

    b A boat with green sails was the only boat to capsize.

       What is the probability that Peter was on a boat that capsized?

    c A boat with straight sails came last.

       What is the probability that Andy was on the boat that came last?

4 Look at the shapes in the grid.

Sumi chooses a shape from the grid at random.
What is the probability of it being:

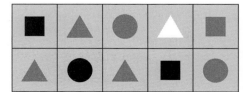

    a a red square   **b** a black square   **c** a square   **d** a red circle
    e a black circle   **f** a circle   **g** a red triangle   **h** a triangle?

5 A number square contains the numbers from 1 to 100.

1	2	3	4	5	6	7	8	9	10
11	12	13	14	15	16	17	18	19	20
21	22	23	24	25	26	27	28	29	30
31	32	33	34	35	36	37	38	39	40
41	42	43	44	45	46	47	48	49	50
51	52	53	54	55	56	57	58	59	60
61	62	63	64	65	66	67	68	69	70
71	72	73	74	75	76	77	78	79	80
81	82	83	84	85	86	87	88	89	90
91	92	93	94	95	96	97	98	99	100

Tom chooses a number from the number square at random. What is the probability of it being:

a a number greater than 50

b a number less than 10

c a square number (1, 4, 9, …)

d a multiple of 5

e a number with at least one 6 in it

f a square number with a nine in it

g an even number with a 7 in it

h an odd number containing a 2

i a number smaller than 23 with a 1 in it?

6 Look at the calendar for the first three months of 2014.

Joy has a birthday in the first three months of the year.

What is the probability of her birthday in 2014 being:

a in February

> There are 28 days in February and 90 days in the first three months of the year.
> So the probability is $\frac{28}{90} = \frac{14}{45}$.

The first one is done for you.

b on a Wednesday **c** on a Monday in January **d** on the 28th

e on 18th February **f** on a March Sunday **g** on a Tuesday **h** on the 1st

i in March **j** at the weekend **k** on a weekday **l** on the 31st?

Activity: Boats

A Draw your own set of 10 boats so that:

  a there are more boats with blue hulls than any other colour

  b there are fewer boats with red hulls than any other colour

  c there are the same number of boats with flags at the top of the mast as boats without flags at the top of the mast.

B You choose one of your boats at random. What is the probability that it will have:

  a blue sails   **b** a red hull

  c a flag at the top of the mast

  d a flag at the top of the mast and a blue hull?

3.4 Using a sample space to calculate probabilities

Learning objective

- To use sample spaces to calculate probabilities

Key word

sample space

To help you work out the probabilities of events happening together you can use tables or diagrams called **sample spaces**.

A sample space is the set of all possible outcomes from a specific event.

Some events are simple, such as rolling a dice.

The sample space is {1, 2, 3, 4, 5, 6}.

Some are more complicated, such as rolling a dice and tossing a coin.

Example 7

Find the probability of getting a head and a six when you roll a dice and toss a coin at the same time.

This sample space shows all the possible outcomes of throwing a coin and a dice together.

	1	2	3	4	5	6
Head	H, 1	H, 2	H, 3	H, 4	H, 5	H, 6
Tail	T, 1	T, 2	T, 3	T, 4	T, 5	T, 6

You can now work out the probability of getting both a head and a 6.

$$P(outcome) = \frac{\text{the number of ways the outcome could occur}}{\text{the total number of possible outcomes}}$$

P(head and a six) = $\frac{1}{12}$

Example 8

Three friends are deciding what to drink with their lunch: tea, coffee or water.

a Draw the sample space of all the different combinations of drinks they could choose.

b What is the probability that any two of the friends have chosen water, but not all three of them?

   a Draw the sample space in a logical order such as shown here.

tea, tea, tea	tea, tea, coffee	tea, tea, water
tea, coffee, coffee	tea, coffee, tea	tea, coffee, water
tea, water, water	tea, water, coffee	tea, water, tea
coffee, coffee, coffee	coffee, coffee, tea	coffee, coffee, water
coffee, tea, tea	coffee, tea, coffee	coffee, tea, water
coffee, water, water	coffee, water, coffee	coffee, water, tea
water, water, water	water, water, coffee	water, water, tea
water, coffee, coffee	water, coffee, tea	water, coffee, water
water, tea, tea	water, tea, coffee	water, tea, water

     This gives every combination of choices from all three friends – there are 27 outcomes altogether.

   b You can find probability that just two of them have chosen water by adding how many outcomes in the sample space have just two waters. The outcomes are:

     tea, water, water   coffee, water, water   water, tea, water   water, coffee, water

     water, water, coffee   water, water, tea

     There are six outcomes.

     So P(two waters) = $\frac{6}{27}$ = $\frac{2}{9}$.

Example 9

In a class of 30 pupils, there are 16 boys. There are 5 pupils in the class who wear glasses and two of them are girls.

What is the probability of choosing a boy from the class who does not wear glasses?

   You could put this information into a table.

	Boy	Girl	Total
Wear glasses	3	2	5
Do not wear glasses	13	12	25
Total	16	14	30

You can now see that:

P(boy not wearing glasses) = $\frac{13}{30}$

You can also show this on a Venn diagram, like this, and use it to help you work out the probability.

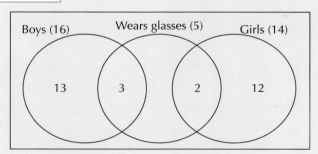

Exercise 3D

1 The sample space for throwing a coin and a dice together is shown.

	1	2	3	4	5	6
Head	H, 1	H, 2	H, 3	H, 4	H, 5	H, 6
Tail	T, 1	T, 2	T, 3	T, 4	T, 5	T, 6

Use the sample space to find the probability of each outcome.

The first one is done for you.

a a 2 and a head   **b** a head and an odd number   **c** a number less than 3 and a tail

> There is one possible way the outcome can occur.
> There are 12 possible outcomes altogether.
>
> P(a 2 and a head) = $\frac{1}{12}$

d a 4 and a tail   **e** a tail and an even number   **f** a head and a number greater than 2

2 The sample space for the three friends choosing their drinks is shown.

tea, tea, tea	tea, tea, coffee	tea, tea, water
tea, coffee, coffee	tea, coffee, tea	tea, coffee, water
tea, water, water	tea, water, coffee	tea, water, tea
coffee, coffee, coffee	coffee, coffee, tea	coffee, coffee, water
coffee, tea, tea	coffee, tea, coffee	coffee, tea, water
coffee, water, water	coffee, water, coffee	coffee, water, tea
water, water, water	water, water, coffee	water, water, tea
water, coffee, coffee	water, coffee, tea	water, coffee, water
water, tea, tea	water, tea, coffee	water, tea, water

Use the sample space to find the probability of each event.

a Only one of them chooses tea.   **b** They all choose different drinks.

c Just two of them choose coffee.   **d** Two of them choose the same drink.

e All three choose the same drink.   **f** None of them choose water.

3 Alice rolls two dice and adds the scores together.

a Copy and complete the sample space of her scores.

	1	2	3	4	5	6
1	2	3				
2	3					

b Use the sample space to work out the probability that the total score is:

i 3   **ii** 6   **iii** 7   **iv** 11   **v** less than 8

vi less than or equal to 7   **vii** greater than or equal to 9

viii odd   **ix** 7 or 1   **x** greater than 4   **xi** greater than 7.

4 At a school disco the flavours of packets of crisps sold are plain, cheese and onion or salt and vinegar.

Serena and Matt each buy a packet of crisps.

a Copy and complete their sample space.

Serena	Matt
plain	plain
plain	cheese and onion

b Use your sample space to work out the probability of:

i Serena choosing plain **ii** Matt choosing plain **iii** both choosing plain

iv Serena choosing plain and Matt choosing salt and vinegar

v Serena choosing salt and vinegar and Matt choosing cheese and onion

vi both choosing the same flavour

vii both not choosing plain

viii both choosing different flavours.

MR **5** A club has equal numbers of men and women. First one member is chosen, then another.

a Write down the four possible combinations that could be chosen.

b Jess says that the probability of choosing two women is $\frac{1}{3}$. She is wrong. Explain why she is wrong.

MR **6** A box contains cubes, spheres and cylinders. Two shapes are chosen at random.

a List the possible outcomes.

b Explain how you could have more of a chance of choosing a cube or a sphere rather than a cylinder.

7 Two dice are rolled and the score is the difference of the two numbers.

a Draw the sample space of the scores.

b What totals are the most likely to occur?

c What would be the probability of rolling a score less than three?

PS **8** Two spinners each showing the numbers 1, 2, 3 and 4 are spun together.

What is the probability that the sum of the numbers spun is less than five?

PS **9** Which one of these outcomes has the greater probability? Explain your answer.

The chance of rolling one dice and getting a six or the chance of rolling two dice and getting a total of six.

10 For a conference break, the kitchen staff made 100 coffees.

70 of them were made with milk, the rest with no milk.

50 of them had sugar in.

There were 12 coffees without milk and sugar.

a Show this information on a Venn diagram.

b Use your Venn diagram to find the probability of picking at random a coffee with milk but no sugar.

Problem solving: Socks

Oliver has a sock draw where all the socks are mixed up. He knows that he has two pairs of blue socks, one pair of white socks and one pair of red socks.

He takes two socks out without looking at what colour they are. What is the probability that these socks are both the same colour?

 Hint The answer is NOT $\frac{1}{3}$.

3.5 Experimental probability

Learning objective

• To calculate probabilities from experiments

Key word

experimental probability

How could you estimate the probability that a train will be late?

Will the train be late again today?

You could keep a record of the number of times that the train arrives late over a period of 10 days, and then use these results to estimate the probability that it will be late in future. The experiment here is observing the train, and the outcome you record is the train being late. The results enable you to find the **experimental probability** of an event.

$$\text{Experimental probability} = \frac{\text{number of trials that produced the outcome}}{\text{total number of trails carried out}}$$

Example 10

An electrician wants to estimate the probability that a new light bulb lasts for less than one month.

He fits 10 new bulbs and two of them fail within 1 month.

What is his estimate of the probability that a new light bulb fails within the month?

2 out of 10 bulbs fail within 1 month, so his experimental probability is $\frac{2}{10} = 0.2$.

Example 11

A dentist keeps a record of the number of fillings the practice gives her patients over a week. Here are her results.

Number of fillings	None	1	More than 1
Number of patients	53	36	11

Estimate the probability that a patient does not need a filling. There are 100 patients altogether.

Experimental probability = $\frac{53}{100}$ = 0.53

Example 12

A company manufactures items for computers. The number of faulty items is recorded in this table.

Number of items produced	Number of faulty items	Experimental probability
100	8	0.08
200	20	
500	45	
1000	82	

a Copy and complete the table.

b Which is the best estimate of the probability of an item being faulty? Explain your answer.

a

Number of items produced	Number of faulty items	Experimental probability
100	8	0.08
200	20	0.1
500	45	0.09
1000	82	0.082

b The best estimate is the last result (0.082), as the experiment is based on more results.

Exercise 3E

1 Asha decides to carry out an experiment to estimate the probability of a slice of bread and butter landing butter side up when it is dropped. She drops a slice of bread and butter 50 times and records the results. She then repeats the experiment several times with different slices of bread and butter. Her results are shown in the table.

Number of times dropped	Number butter side up
50	19
100	29
150	45
200	59
250	72

a Would you say that there is a greater chance of a slice of bread and butter landing buttered side up or down? Explain your answer.

b Explain which result is the most reliable.

c Estimate the probability of a slice of bread and butter landing butter side up.

d How could Asha improve the experiment?

 2 Ben wants to test a dice to see if it is biased. He rolls the dice 50 times. His results are shown in the table.

Score	1	2	3	4	5	6
Frequency	5	11	8	8	10	8

a Do you think the dice is biased? Give a reason for your answer.

b How could Ben improve the experiment?

c From the results, estimate his probability of rolling 1.

d From the results, estimate his probability of rolling 5 or 6.

 3 Michelle's train often seemed to get her to work late.

Over 5 weeks, she kept a count of how many times it made her late.

After	Days	Number of times late	Probability
1 week	5	3	
2 weeks	10	5	
3 weeks	15	6	
4 weeks	20	7	
5 weeks	25	9	

a Work out the probabilities after each week, of the train making her late for work.

b What would you say is the probability of the train making Michelle late for work?

c Explain your answer to part **b**.

4 Storme helped her mum at a fairground hoopla stall.

One evening she decided to count each hour how many throws had been made, how many hoops missed altogether, how many went over the item but not over the base, and how many went over the base and won a prize.

Her results are shown in this table.

Time pm	Number of throws	Number of times missed	Number of times over the prize but not the base	Number of times over the base and a prize won
7–8	45	28	16	1
8–9	75	42	31	2
9–10	90	49	39	2
10–11	80	36	41	3

a Work out the probability that someone won a prize between 7 pm and 8 pm.

b What is the probability, over the whole evening, of missing?

c What is the probability, over the whole evening of someone winning a prize?

d Explain how Storme could get a more accurate probability of the chance of someone winning a prize.

Problem solving: Rolling two dice

Use two dice to see what the probability is of:

A rolling a total higher than seven   **B** rolling an odd total.

Ready to progress?

I understand about chance.

I can use a probability scale from 0 to 1.
I can find the experimental probability from a simple experiment.
I understand that different outcomes may result from repeating an experiment.

I can use the probability of an event to calculate the probability that the event does not happen.

Review questions

1 Make three copies of the diagram of a spinner.

On each copy of the spinner, write five numbers to make the statements below correct.

a It is certain that you will get a number less than 5.

b It is more likely that you will get an odd number than an even number.

c It is impossible that you will get a multiple of 3.

2 Owen buys a box of cat food that contains 10 packets of different varieties. The table shows the different varieties in the box.

a Owen takes out a packet at random from the box. What is the probability that it will be tuna?

b Owen's cat eats all the packets of tuna. He takes a packet at random from the ones left in the box. What is the probability that it will be salmon?

Variety	Number of packets
Cod	3
Salmon	2
Trout	2
Tuna	3

3 Kieron chooses a number, at random, from the list of numbers.

2   3   9   6   5   10   4   13   25   8

What is the probability that the number he chooses is:

a a number that divides exactly into 10

b a square number

c a number that divides exactly into 18 and 36

d a number that will only divide exactly by itself and one?

4 Emma draws all the triangles she can where one of the angles is always 100° smaller than another angle and all the angles are multiples of 10°.

a Sketch all the three different triangles Emma could have drawn, showing the angles.

She drew all her triangles a different colour. One of the colours was blue.

b What is the probability that her blue triangle has:

i an angle of 20°   **ii** a right angle in it?

c One of the other triangles was red. What is the probability that this triangle was obtuse-angled?

5 In each box of tea bags there is a free gift of a card.

You cannot tell which card will be in a box. Each card is equally likely.

There are four different cards showing a picture of a koala, a toucan, a sea horse and a camel.

   a Billy needs the card with the koala.

     His sister Vicky needs the cards with the toucan and the sea horse.

     Their dad buys one box of teabags.

     i What is the probability that the card is one that Billy needs?

     ii What is the probability that the card is one that Vicky needs?

   b Then their dad opens the box. He tells them the card is not a card with a koala.

     i Now what is the probability the card is one that Billy needs?

     ii What is the probability that the card is one that Vicky needs?

 6 a Sophia puts three blue marbles and one white marble into a bag.

     She takes one marble out of the bag without looking.

     What is the probability that the marble will be blue?

   b Sophia puts the marble back in the bag and then puts more white marbles into the bag.

     She takes one marble out of the bag without looking.

     The probability that the marble will be blue is now $\frac{1}{2}$.

     What is the least number of marbles Sophia put in the bag and what colours are they?

     Explain your answer.

7 At the local newsagent the flavours of packets of crisps sold are plain, chicken or salt and vinegar.

Scott and Jenny each buy a packet of crisps.

   a Copy and complete their sample space.

Scott	Jenny
plain	plain
plain	chicken

   b Use your sample space to work out the probability of:

     i Scott choosing chicken   **ii** Jenny choosing chicken   **iii** both choosing chicken

     iv Scott choosing plain and Jenny choosing salt and vinegar

     v Scott choosing salt and vinegar and Jenny choosing chicken

     vi both choosing the same flavour   **vii** both not choosing plain

     viii both choosing different flavours.

Financial skills
Fun in the fairground

Hoopla

The fair has come to town.

You can buy five hoops for £1.50.

You win a prize by throwing a hoop over that prize, but it must also go over the base that the prize is standing on!
Ben spent some time watching people have a go at this stall and started to count how many goes they had and how many times someone won.

This table shows his results.

Prize	Number of throws	Number of wins
Watch	320	1
£10 note	240	4
£1 coin	80	2

Use the information on **Hoopla** to answer these questions.

1 How many sets of 5 hoops thrown did Ben observe?

2 What income would these throws have made for the stall?

3 From the results shown, what is the probability of someone aiming for and winning a:

    a £1 coin   **b** £10 note   **c** watch?

4 What would you say is the chance of someone winning a prize with:

    a one hoop   **b** five hoops?

5 After watching this, Ben decided to try for a £10 note.
   He bought 25 hoops and all his throws were aimed at the £10 note.

    a How much did this cost him?

    b What is his probability of winning a £10 note?

Hook a duck

On this stall, plastic ducks float in a moat around a central stall. Each duck has a number written on its underside, which cannot be seen until the duck is caught, by means of a hook on a stick. The number is checked by the stall holder.

If the number on the duck is:

1 you win a lollipop   2 you win a yo-yo   5 you win a cuddly toy.

Each time a duck is hooked, it is replaced in the water. Cindy, the stall holder, set up the stall one week with:

🦆 45 plastic ducks

🦆 only one of which had the number 5 underneath

🦆 nine had the number 2 underneath

🦆 all the rest had a number 1 underneath.

Cindy charged 50p for one stick, to hook up just one duck.

Use the information about **Hook a duck** to answer these questions.

6 What is the probability of hooking the number:

   a 1   **b** 2   **c** 5?

7 What is the probability of winning:

   a a cuddly toy   **b** a yo-yo   **c** a lollipop?

8 What is the probability of winning anything other than a lollipop?

9 Tom spent £2 at the stall so that his little sister, Julie, could hook some ducks.

   a How many ducks could Julie hook with the £2?

   b What is the likelihood that, after four goes, Julie has won:

      i a cuddly toy   **ii** a yo-yo   **iii** a lollipop?

   In each case, choose from:
   impossible, very unlikely, unlikely, evens, likely, very likely, certain.

10 Before lunch on Sunday, Cindy took £120 from the stall.

   a How many ducks had been hooked that morning?

   b How many cuddly toys would you expect Cindy to have given away that morning?

   c How many yo-yos would you expect Cindy to have given away that morning?

4

Percentages

This chapter is going to show you:

- how to write one value as a percentage of another value
- how to calculate a percentage increase or decrease
- how to write a change of value as a percentage increase or decrease.

You should already know:

- the equivalence between fractions, decimals and percentages
- how to calculate a percentage of a quantity with or without a calculator.

About this chapter

Pay rises are often given as percentages. A pay rise of 4% is not the same amount of extra money for everyone. If you earn £200 a week you will earn another £8 a week. If you earn £1000 a week you will earn another £40 a week.

When you are comparing increases and decreases it is best to use percentages. For example an increase of £2000 in a price sounds like a lot of money but it depends on the original price. If the price of a car increases from £5000 to £7000 that is a very big rise. If the price of a house increases from £200 000 to £202 000 it does not make much difference. You can see this more clearly by writing the increases as percentages. The price of the car has increased by 40%. The price of the house has only increased by 1%.

4.1 Calculating percentages

Learning objective

- To write one quantity as a percentage of another

Key words

| percentage | unitary method |

You should be able to calculate a **percentage** of a quantity, using a calculator where necessary. Example 1 reminds you how to do that.

Often we know two values and we want to write one as a percentage of the other. This section will show you how to do that.

Example 1

Lucy has £240. She gives 15% to a charity. How much is that?

Here are two different methods.

Method 1

1% of £240 = £2.40	To find 1% of £240, divide it by 100.
15% of £240 = 15 × £2.40 = £36.00	15% is 15 × 1%

This is called a **unitary method** because you start by finding one per cent.

Method 2

15% = 0.15	To write a percentage as a decimal, divide it by 100.
15% of £240 = 0.15 × £240 = £36.00	Multiply 240 by the decimal.

Both methods give the same answer. You can use either method.

Example 2

Jon has a 500 g bag of flour. He uses 150 g in a recipe. What percentage of the flour does he use?

Write 150 as a fraction of 500 and change the fraction to a percentage.

$$\frac{150}{500} = 150 \div 500 = 0.3$$

0.3 = 0.3 × 100% = 30%   Multiply the decimal by 100 to find the percentage.

Example 3

Petra has £63.50 and spends £53.34. What percentage of her money does she spend?

You can use the same method, even though the numbers are decimals.

The fraction is $\frac{53.34}{63.5} = 53.34 \div 63.5 = 0.84$.

0.84 × 100 = 84%   Multiply the decimal by 100 to find the percentage.

Exercise 4A

1 Find 1% of:
  a £37   **b** £85   **c** £324   **d** £6200   **e** £93 000.

2 Find 1% of:
  a 700 cm   **b** 325 km   **c** 420 kg   **d** 34 000 people   **e** 6720 m.

3 Work out 45% of:
  a £80   **b** £360   **c** £9   **d** 140 kg   **e** 2500 m.

4 Work out each percentage of £34.50.
  a 20%   **b** 64%   **c** 8%   **d** 92%   **e** 110%

5 This line shows percentages of £64.

Use it to find each percentage of £64.

  a 10%   **b** 30%   **c** 80%   **d** 40%   **e** 70%   **f** 5%

(MR) 6 Show that 16% of £83 is the same as 83% of £16.

7 Work out these values.
  a 35% of £84   **b** 19% of £124   **c** 62% of 40 kg   **d** 8% of 172 m
  e 95% of £2300   **f** 22% of 22 km   **g** 130% of £60   **h** 12.5% of £560

8 Here are some test marks. Write them as percentages.
  a 24 out of 100   **b** 24 out of 200   **c** 24 out of 50   **d** 24 out of 75
  e 24 out of 40   **f** 24 out of 30   **g** 24 out of 25   **h** 24 out of 35

9 Write these amounts of money as percentages.
  a £3 out of £60   **b** £12 out of £60   **c** £27 out of £60
  d £45 out of £60   **e** £54 out of £60   **f** £37 out of £60

10 Write these quantities as percentages.
  a 15 kg out of 24 kg   **b** 27 kg out of 90 kg   **c** 64 kg out of 80 kg
  d 324 g out of 480 g   **e** 650 g out of 2000 g   **f** 900 g out of 1250 g

11 Write these quantities as percentages.
  a 25 ml out of 200 ml   **b** 24 cm out of 80 cm   **c** 324 km out of 400 km
  d 231 days out of 300 days   **e** 72 years out of 320 years   **f** 3 m out of 60 m

(PS) 12 Candy is running in a 5 kilometre race. This is 12.5 laps of a running track.
  What percentage of the race has she completed after she has run:
  a 1 km   **b** 4.5 km   **c** 5 laps   **d** 10 laps   **e** 1 lap?

13 Write these amounts as percentages. Give your answer to the nearest whole number.
  a £33 out of £72   **b** £65 out of £264   **c** £6.50 out of £74
  d £19 out of £143   **e** £73 out of £84   **f** £126 out of £421

(FS) **14** Jason earns £620. He pays £76 income tax.

What percentage of his pay is the income tax?

(FS) **15** Peter has £680 at the start of the week.

a He spends £185 on his rent. Show that this is 27% of his money, to the nearest whole number.

b Copy and complete this table showing how he spent his money.

Item	Amount	Percentage
Rent	£185	27%
Food	£128	
Clothes	£49.50	
Travel	£32.60	

 16 There are 40 people in a room.

a 10 people enter the room. Cassie says: 'The number of people in the room has increased by 25%'. Is Cassie correct? Justify your answer.

b 10 more people enter the room. Cassie says: 'The number of people in the room has increased by another 20%'. Is Cassie correct? Justify your answer.

Challenge: What is in the waste?

The Watson family weigh the waste they throw away. Here are the results.

Type of waste	Mass (g)
Kitchen scraps	1050
Plastics	470
Card or paper	620
Other	820
Total	2960

A Work out the percentage of each type of waste that the family throw away.

B Draw a suitable chart to show your percentages.

4.2 Calculating the result of a percentage change

Learning objective

• To calculate the result of a percentage increase or decrease

You have already solved problems involving percentage increases and decreases. This section will show you an efficient way to do it with a calculator.

Example 4

The price of an article before tax is £164.

20% tax must be added.

Work out the price including tax.

£164 is 100%. You need to add 20%.

You want 100% + 20% = 120% all together.

1% of 164 = 1.64   That is 164 ÷ 100.

120% of 164 = 1.64 × 120 = 196.8

The price including the tax is £196.80.   You need the extra 0 because it is pounds.

With practice, you can do this calculation all together.

164 ÷ 100 × 120 = 196.8   Check you can do this on your calculator.

Example 5

The price of a pair of shoes is £64.50.

In a sale, the price is reduced by 30%.

Work out the sale price.

100% = 64.5

This time you need to subtract 30%.

You want 100% − 30% = 70%. You can use the unitary method and start with 1%.

1% = 64.5 ÷ 100 = 0.645

70% = 0.645 × 70 = 45.15

The sale price is £45.15.

Check that you can get the answer in one set of calculations: 64.5 ÷ 100 × 70 = 45.15.

Exercise 4B

(FS) 1 The price of a food mixer is £74.00.

The price increases by 10%.

    a What is 1% of £74.00?   **b** Work out the price after the increase.

(FS) 2 The price of a TV is £385. The price is reduced by 35% in a sale.

    a What is 100% − 35%?   **b** Work out 1% of £385.

    c Calculate the sale price.

(FS) 3 Increase each price by 10%.

    a £17.00   **b** £120.00   **c** £42.50   **d** £243.20

(FS) 4 Decrease each price by 10%.

    a £42.00   **b** £270.00   **c** £23.90   **d** £85.30

5 Increase each mass by 35%.

    a 60 kg   **b** 9 kg   **c** 148 kg   **d** 88 kg

6 Decrease each mass by 35%.

   a 90 kg   **b** 13 kg   **c** 580 kg   **d** 7 kg

7 Increase each measurement by 60%.

   a 320 m   **b** 16 km   **c** 70 cm^2 **d** 6300 litres

8 Decrease each measurement by 60%.

   a 7 km   **b** 84 cm   **c** 170 m^2 **d** 580 ml

(FS) 9 20% tax must be added to these prices. Work out the prices including tax.

   a £63.50   **b** £6.75   **c** £279   **d** £141.60

(FS) 10 Increase £280 by:

   a 5%   **b** 45%   **c** 85%   **d** 95%   **e** 120%.

11 Increase 62 kg by:

   a 7%   **b** 38%   **c** 83%   **d** 105%   **e** 136%.

(FS) 12 These prices will be reduced by 40% in a sale. Work out the sale prices.

   a £690   **b** £72.50   **c** £19.90   **d** £5320

(FS) 13 The price of a holiday is £2590. Work out the new price when it is reduced by:

   a 5%   **b** 12%   **c** 18%   **d** 35%.

(FS) 14 The price of a coat is £130.

   a Copy and complete the table to show the price after different reductions.

Full price	20% off	40% off	60% off	80% off
£130				

   b Copy and complete these sentences:

     i The price with 80% off is half the price with ....% off.

     ii The price with 60% off is half the price with ....% off.

Financial skills: Flow chart

A Copy this flow chart and fill in the missing numbers.

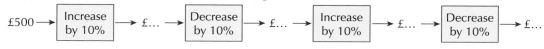

£500 → [Increase by 10%] → £... → [Decrease by 10%] → £... → [Increase by 10%] → £... → [Decrease by 10%] → £...

B Repeat part **A**, but replace 10% by 20% in each box.

4.3 Calculating a percentage change

Learning objective

• To work out a change of value as a percentage increase or decrease

In the last section you found the result of increasing or decreasing a value by a given percentage.

Sometimes you will know the increase or decrease in value and you want to work out the percentage change from the original value.

Example 6

The number of students in a college increases from 750 to 795.

Work out the percentage increase.

The increase is $795 - 750 = 45$.

Write this as a percentage of the original number (750).

$\frac{45}{750} = 45 \div 750 = 0.06$ Change the fraction to a decimal by dividing by 750.

$0.06 \times 100 = 6\%$

The increase is 6%.

Example 7

In a sale the price of a washing machine is reduced from £449 to £389.

Work out the percentage reduction.

The reduction is $449 - 389 = 60$.

Write this as a percentage of the original price (£449).

$\frac{60}{449} = 60 \div 449 = 0.1336...$ Leave this number on your calculator and use it for the next calculation.

$0.1336... \times 100 = 13.36...\% = 13\%$ to the nearest whole number.

Do not round off until the end of the calculation.

Exercise 4C

 1 The cost a train ticket is £35.00.

The price increases by £2.80.

Work out the percentage increase.

 2 The cost of hiring a van for a day is £76.

The price is reduced by £11.40.

Work out the percentage decrease.

(FS) 3 The price of a computer increases from £550 to £616.

    a What is the increase in price?

    b Work out the increase as a percentage of the original price.

4 The mass of a child increases from 7.0 kg to 9.8 kg.

    a What is the increase in mass?   **b** What is the percentage increase?

5 The number of trees in a plantation decreases from 120 to 102.

    a What is the decrease in the number of trees?

    b What is the decrease as a percentage of the original number of trees?

6 The mass of a person decreases from 60.0 kg to 57.6 kg.

    a What is the decrease in mass?   **b** Calculate the percentage decrease.

7   The number of people employed by a company increases from 240 to 324.

    Work out the percentage increase.

FS 8   The cost of a second-hand car is reduced from £3700 to £3293.

    Calculate the percentage reduction.

FS 9   The cost of posting a parcel increases from £7.25 to £8.12. Work out the percentage increase.

10   The mass of a machine is reduced from 645.0 kg to 464.4 kg. Work out the percentage reduction in mass.

11   Work out these percentage increases. Give your answers to the nearest whole number.

    a   A price increases from £324 to £399.

    b   A cost increases from £37.50 to £49.80.

    c   A mass increases from 7.9 kg to 12.8 kg.

    d   A volume increases from 293 cm^3 to 402 cm^3.

FS 12   All of these prices are reduced by £50. Work out the percentage change in each case. Give your answers to the nearest whole number.

    a   camera £375   b   bike £485   c   mobile phone £139   d   sofa £849

13   Work out the percentage change in each case. Give your answers correct to the nearest whole number.

    a   A population increases from 45 000 to 53 000.

    b   A price is reduced from £95.50 to £63.90.

    c   A fuel bill increases from £217 to £269.

    d   A journey time is reduced from 23 hours to 16 hours.

Financial skills: Pay rises

A   Three people are given a 4% pay rise.

Copy and complete this table to show the new weekly pay for each person.

Name	Old weekly pay	Pay rise	New weekly pay
Ms Scarlet	£420.50		
Mrs White	£842.30		
Mr Black	£625.70		

B   Three people are given a weekly pay rise of £20. Copy this table and complete it to show the percentage rise for each person.

Name	Old weekly pay	New weekly pay	Percentage rise
Ms Green	£753.40		
Mr Grey	£357.65		
Mr Brown	£904.21		

C   When a group of people are given a pay rise it is usually a percentage increase (such as 4%) rather than the same amount for everyone (such as £20).

Do you think it would be better to give everyone the same percentage or the same amount? Give a reason for your answer.

Ready to progress?

 I can calculate a percentage of a particular amount.

 I know how to write one value as a percentage of another value.
I can work out the result of a percentage increase or decrease.
I can write a change of value as a percentage increase or decrease.

Review questions

1   Work these out.

    a   17% of £185   b   7% of £185   c   97% of £185   d   117% of £185

2   a   22 boys out of a group of 50 are over 1.50 m tall. What percentage is that?

    b   22 girls out of a group of 40 are over 1.50 m tall. What percentage is that?

3   A holiday costs £1600. Sally pays a deposit of £480. What percentage is that?

4   Al looks in a shop window. He sees this sign.

THIS WEEK ONLY
TV
WAS £450
NOW £406

Midi HiFi system
WAS £270 NOW £216

    a   What is the percentage reduction in the price of the TV?

    b   What is the percentage reduction in the price of the midi hi-fi?

5   All the workers at a company have been given a 4% pay rise. How much will each of these workers' pay actually increase?

    a   David earns £480 a week.

    b   Hanif earns £560 a week.

    c   Chris earns £1240 a month.

    d   Julita earns £1810 a month.

6 a   It rained on 10 days out of the 28 days in February. Write that as a percentage, to the nearest whole number.

   b   In July it rained on 6 days. What percentage of the days in July was that?

 7   There are 52 men and 28 women in a meeting. What percentage of the people are women?

 8   The price of a car repair is £275. Tax (VAT) of 20% must be added to this. Calculate the price including tax.

9   Here are three test marks.

   a   Write each mark as a percentage.

   b   By comparing the percentages, decide which was the best mark.

> English: 43 out of 60
> Maths: 63 out of 80
> Science: 87 out of 120

 10 Here is some data about the children in a breakfast club.

Age	Boys	Girls
Under 9 years old	12	21
9 or over	13	15
Total	25	36

   a   What percentage of the boys are under 9 years old?

   b   What percentage of the girls are aged 9 or over?

11 The cost of petrol is £1.32 per litre.

   The price is reduced by 6%. Work out the new price. Give your answer to the nearest penny.

 12 The population of a town in 1970 was 35 000.

   a   Between 1970 and 1990 it increased by 20%. Work out the population in 1990.

   b   Between 1990 and 2010 it decreased by 20%. Work out the population in 2010.

 13 James is buying a car for £7500.

   He pays a deposit of £2400.

   a   What percentage of the price is the deposit?

   b   What percentage does he still have to pay?

 14 The cost of a train ticket is £38.40.
   Prices are increased by 4%.
   Work out the new price.

 15 The original price of a phone is £159.
   Work out the sale price.

SALE
Prices down
by 60%

Challenge
Changes in population

1 In 1970 the UK population was 55.7 million.
   In 2010 the UK population was 62.3 million.

   a Work out the increase in population from 1970 to 2010.

   b Work out the percentage increase from 1970 to 2010.

2 In 1970 the population of Australia was 12.5 million.
   In 2010 the population of Australia was 22.1 million.

   a Work out the increase in population from 1970 to 2010.

   b Work out the percentage increase from 1970 to 2010.

3 a Which country, UK or Australia, had a bigger population increase from 1970 to 2010?

   b Which country, UK or Australia, had a bigger percentage change from 1970 to 2010?

4 This table shows the populations of the UK and Australia in different years. The numbers are in millions.

	1970	1980	1990	2000	2010
UK	55.7	56.3	57.3	58.9	62.3
Australia	12.5	14.7	17.1	19.2	22.1

   a Draw a graph to show the changes in the UK population. Draw axes like this.

   Join the points with a smooth curve.

   b On the same axes show the change in the Australian population.

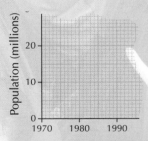

5 The world population in 1980 was 4.44 billion. A billion is 1000 million so 4.44 billion is 4440 million.

a Write 4.44 billion in figures.

b The world population in 1990 was 5.26 billion. Write this in figures.

c Work out the increase in the world population between 1980 and 1990.

d Work out the percentage increase in the world population between 1980 and 1990.

e This table shows the world population at three different times.

Year	1980	1990	2000
World population (billions)	4.44	5.26	6.07

You have already worked out the percentage increase from 1980 to 1990.
Work out the percentage increase from 1990 to 2000.

6 This table shows the populations of different parts of the world in 2008 and estimated populations in 2025. The numbers are in millions.

	2008	2025
Asia	2183	2693
Africa	984	1365
Europe	603	659
South America	462	550
North America	444	514

a Write the population of Asia in 2008 in figures.

b Write the predicted population of Asia in 2025 in figures.

c Show that the predicted percentage increase of the population of Asia from 2008 to 2025 is 23%, to the nearest whole number.

d Copy and complete this table to show the predicted percentage changes in populations.

	Predicted change 2008 to 2025
Asia	23%
Africa	
Europe	
South America	
North America	

e

The region with the largest population will have the largest percentage change.'

Is this correct? Justify your answer.

5

Sequences

This chapter is going to show you:

- how to use the special sequence of Fibonacci numbers
- how to use algebra to represent function machines
- how to use the nth term for sequences.

You should already know:

- the sequence of square numbers
- the sequence of triangular numbers
- how to use function machines
- that letters can be used for numbers
- how to substitute numbers into an algebraic expression.

About this chapter

The Fibonacci numbers are nature's numbering system. They appear everywhere in nature, from the leaf arrangement in plants, to the pattern of the florets of a flower, the arrangement of the twists on a pinecone or the scales of a pineapple. On many plants, the number of petals is a Fibonacci number: lilies and irises have 3 petals, buttercups have 5 petals, delphiniums have 8 petals, corn marigolds have 13 petals and asters have 21 petals, whereas daisies can be found with 34, 55 or even 89 petals. The Fibonacci numbers are therefore applicable to the growth of every living thing, including a single cell, a grain of wheat, a hive of bees – and even how rabbits breed.

5.1 The Fibonacci sequence

Learning objective

* To know and understand the Fibonacci sequence

Key word

Fibonacci sequence

Look at this sequence of numbers:

1, 1, 2, 3, 5, 8, 13, ...

Can you see a pattern in the sequence? It is called the **Fibonacci sequence**.

Each term after the first two is the sum of the previous 2 terms.

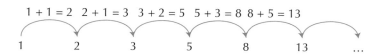

An Italian mathematician, Leonardo Fibonacci, first wrote about this sequence in the thirteenth century.

Example 1

Work out the next two terms of the Fibonacci sequence.

13 + 8 = 21, 21 + 13 = 34

So the next two terms are 21 and 34.

Exercise 5A

1. Write out the Fibonacci sequence up to the 12th term.

2. Write down the numbers from your list that are also square numbers.

3. **a** This sequence is the first five terms of the sequence of triangular numbers.

   1, 3, 6, 10, 15

   Explain the term-to-term rule for continuing the sequence.

   b Write out the sequence of triangular numbers up to the 10th term.

   c Which numbers in your list of triangular numbers are also in the Fibonacci sequence?

4. Work out the next four terms of each sequence.

   a 2, 2, 4, 6, 10, 16, ...   **b** 3, 3, 6, 9, 15, 24, ...

   c 1, 3, 4, 7, 11, 18, ...   **d** 2, 4, 6, 10, 16, 26, ...,

   > **Hint** Think how the Fibonacci sequence works.

5. Look at your list of Fibonacci numbers. Which ones are prime numbers?

 6. **a** The sum of the first three terms of the Fibonacci sequence is:

   1 + 1 + 2 = 4.

   i Work out the sum of the first four terms of the Fibonacci sequence.

   ii Work out the sum of the first five terms of the Fibonacci sequence.

   iii Work out the sum of the first six terms of the Fibonacci sequence.

   b What is the sum of the first ten terms of the Fibonacci sequence?

   Explain how you obtained your answer.

(PS) **7** Write down three consecutive terms from the Fibonacci sequence.

Square the middle number.

Multiply the first and last numbers.

> **Hint** Consecutive means following on.

Write down the difference between these two numbers.

For example: 5, 8, 13

$8^2 = 64$ and $5 \times 13 = 65$, so the difference $= 65 - 64 = 1$.

a Repeat this rule for the three terms 2, 3, 5.

What do you notice?

b Repeat this rule for another two sets of three consecutive terms from the Fibonacci sequence.

Problem solving: Dividing Fibonacci terms

A Divide each term in the Fibonacci sequence by the one before it.

Copy and complete this table.

Term	Previous term	Division	Answer
1	1	1 ÷ 1	1
2	1	2 ÷ 1	2
3	2	3 ÷ 2	1.5
5	3	5 ÷ 3	1.666 66...
8	5	8 ÷ 5	
13	8		
21			

> **Hint** If the answer is a decimal, write down the first five decimal places from your calculator display.

B Explain what is happening.

5.2 Algebra and function machines

Learning objective

• To use algebra with function machines

> **Key word**
> algebraic function machine

You have already met function machines.

This function machine shows the output numbers for the rule 'add 6'.

The rule in the operation box of a function machine can be written using the letter n, where n stands for any number.

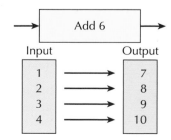

You can now complete this function machine as an **algebraic function machine**.

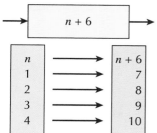

Example 2

Complete the output box for this function machine.

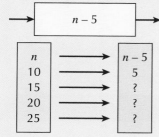

The output box is:

5
10
15
20

Example 3

Complete the output box for this function machine.

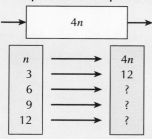

Remember that $4n$ means $4 \times n$.

The output box is:

12
24
36
48

Example 4

Complete the output box for this function machine.

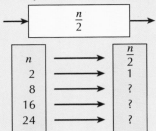

Remember that $\frac{n}{2}$ means $n \div 2$.

The output box is:

1
4
8
12

Example 5

Write down the algebraic rule for this function machine.

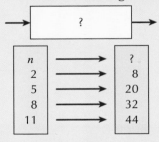

The algebraic rule is: | $4n$ |

Example 6

Complete the output box for this function machine.

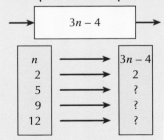

The output box is:

2
11
23
32

Exercise 5B

1. Write down the outputs for each function machine.

a

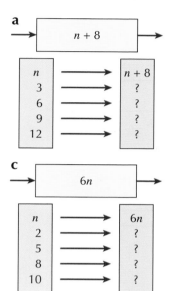

$n + 8$

n	$n + 8$
3	?
6	?
9	?
12	?

b

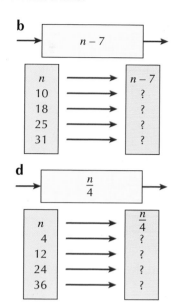

$n - 7$

n	$n - 7$
10	?
18	?
25	?
31	?

c

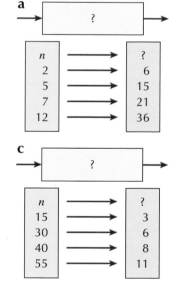

$6n$

n	$6n$
2	?
5	?
8	?
10	?

d

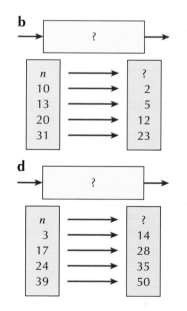

$\dfrac{n}{4}$

n	$\dfrac{n}{4}$
4	?
12	?
24	?
36	?

2. Write down the algebraic rule for each function machine.

a

?

n	?
2	6
5	15
7	21
12	36

b

?

n	?
10	2
13	5
20	12
31	23

c

?

n	?
15	3
30	6
40	8
55	11

d

?

n	?
3	14
17	28
24	35
39	50

(MR) **3** Write down the inputs for each function machine.

a
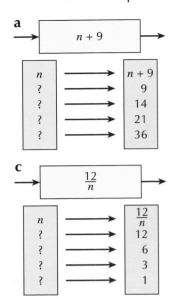

b
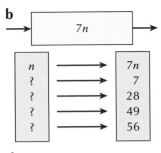
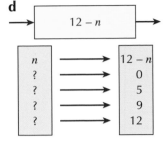

c
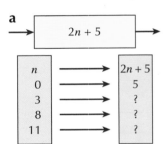

d
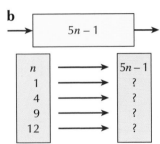

4 Write down the outputs for each function machine.

a

n	$2n + 5$
0	5
3	?
8	?
11	?

b

n	$5n - 1$
1	?
4	?
9	?
12	?

5 Write down the outputs for each function machine.

a
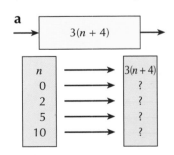

b
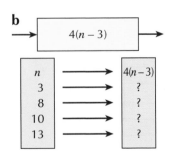

(PS) **6** Write down the inputs for each function machine.

a
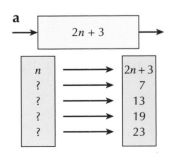

b
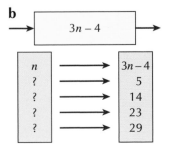

Problem solving: Working out algebraic rules

Work out the algebraic rule for each function machine.

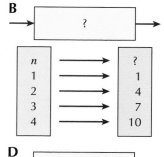

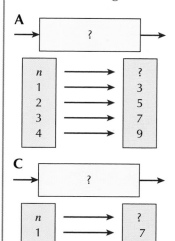

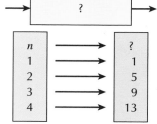

5.3 The nth term of a sequence

Learning objective

• To use the nth term of a sequence

Key words

algebraic expression

nth term

Here is a sequence of numbers.

3, 7, 11, 15, 19, ...

The first term is 3 and the term-to-term rule is 'add 4'.

So the sixth term is 23, the seventh term is 27 and so on.

How do you work out the 10th term?

You could carry on using the term-to-term rule:

the eighth term is 31, the ninth term is 35, so the 10th term is 39.

How would you work out the 50th term?

You can describe sequences by giving a rule to work out any term. This rule is called the **nth term** rule and is an **algebraic expression.**

For the above sequence the nth term is given by the expression $4n - 1$.

You can check the rule for the sixth term: when $n = 6$, the sixth term is $4 \times 6 - 1 = 23$. This is the same answer as you get by using the term-to-term rule.

So the 50th term $= 4 \times 50 - 1 = 199$.

The nth term of the sequence 8, 13, 18, 23, 28, ... is given by the expression $5n + 3$.

a Show this is true for the first three terms.

b Use the rule to work out the 50th term of the sequence.

a When $n = 1$: $5 \times 1 + 3 = 5 + 3 = 8$ true

    When $n = 2$: $5 \times 2 + 3 = 10 + 3 = 13$ true

    When $n = 3$: $5 \times 3 + 3 = 15 + 3 = 18$ true

b When $n = 50$: $5 \times 50 + 3 = 250 + 3 = 253$

The nth term of a sequence is given by the expression $8n - 5$.

a Work out the first three terms of the sequence.

b Work out: **i**   the 8th term   **ii**   the 12th term   **iii**   the 50th term.

a When $n = 1$: $8 \times 1 - 5 = 8 - 5 = 3$

    When $n = 2$: $8 \times 2 - 5 = 16 - 5 = 11$

    When $n = 3$: $8 \times 3 - 5 = 24 - 5 = 19$

    So, the first three terms of the sequence are 3, 11, 19.

b i   8th term $= 8 \times 8 - 5 = 64 - 5 = 59$

ii   12th term $= 8 \times 12 - 5 = 96 - 5 = 91$

iii   50th term $= 8 \times 50 - 5 = 400 - 5 = 395$

Exercise 5C

1 Work out:

    i   the first three terms   **ii**   the 10th term   **iii**   the 20th term

    for the following sequences whose nth term is given by:

    a $2n + 5$ **b** $3n + 2$ **c** $4n + 5$ **d** $10n + 1$ **e** $7n + 8$.

2 Work out:

    i   the first three terms   **ii**   the 8th term   **iii**   the 50th term

    for the following sequences whose nth term is given by:

    a $4n - 1$ **b** $5n - 3$ **c** $7n - 1$ **d** $6n - 6$ **e** $9n - 4$.

3 Work out the first five terms for the following sequences whose nth term is given by:

    a $3n$ **b** $5n$ **c** $7n$ **d** $8n$ **e** $10n$.

    Write down what you notice.

MR **4** Work out the missing number for the *n*th term in each sequence.

   a 3, 5, 7, 9, 11, ...; the *n*th term is 2*n* + ?

   b 8, 13, 18, 23, 28, ...; the *n*th term is 5*n* + ?

   c 2, 6, 10, 14, 18, ...; the *n*th term is 4*n* − ?

   d 2, 9, 16, 23, 30, ...; the *n*th term is 7*n* − ?

5 Work out the first five terms for the following sequences whose *n*th term is given by:

   a 20 − *n*   **b** 18 − 3*n*   **c** 10 − 2*n*   **d** 30 − 5*n*.

PS **6** Look at these patterns, made from matchsticks.

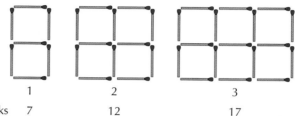

Pattern number	1	2	3
Number of matchsticks	7	12	17

   a How many matchsticks are there in the fourth pattern?

   b How many matchsticks are there in the fifth pattern?

   c Write down the number of matchsticks in the *n*th pattern.

   d How many matchsticks are there in the 50th pattern?

> **Hint** This is the *n*th term.

Problem solving: Matching sequences to their *n*th term

Here are six sequences and a list of expressions for the *n*th term.

6*n* + 1, 4*n* − 2, 3*n* + 5, 6*n* − 2, 5*n* − 1, 4*n* + 4

Match each sequence to its *n*th term. The first one has been done for you.

A 8, 11, 14, 17, 20, ..., *n*th term = 3*n* + 5   **B** 8, 12, 16, 20, 24, ..., *n*th term =

C 7, 13, 19, 25, 31, ..., *n*th term =   **D** 4, 9, 14, 19, 24, ..., *n*th term =

E 2, 6, 10, 14, 18, ..., *n*th term =   **F** 4, 10, 16, 22, 28, ..., *n*th term =

Ready to progress?

I can write out the terms of the Fibonacci sequence.

I can solve problems using the Fibonacci sequence.
I can use algebra with function machines.
I know how to use the nth term of a sequence.

Review questions

1 Write down the outputs for each function machine.

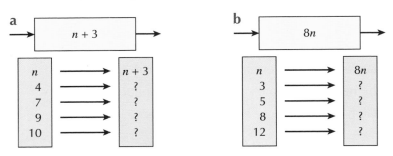

a

n		$n + 3$
4	⟶	?
7	⟶	?
9	⟶	?
10	⟶	?

b

n		$8n$
3	⟶	?
5	⟶	?
8	⟶	?
12	⟶	?

2 Write down the algebraic rule for each function machine.

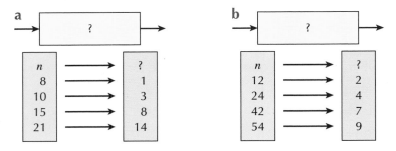

a

n		?
8	⟶	1
10	⟶	3
15	⟶	8
21	⟶	14

b

n		?
12	⟶	2
24	⟶	4
42	⟶	7
54	⟶	9

3 Write down the outputs for each function machine.

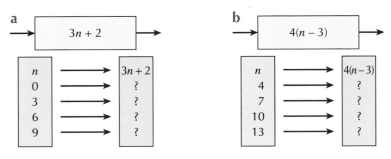

a

n		$3n + 2$
0	⟶	?
3	⟶	?
6	⟶	?
9	⟶	?

b

n		$4(n - 3)$
4	⟶	?
7	⟶	?
10	⟶	?
13	⟶	?

4 Work out: **i** the first three terms   **ii** the 10th term   **iii** the 40th term

for the sequences whose nth term is given by:

a $3n + 4$ **b** $5n + 4$ **c** $4n - 3$ **d** $3n - 1$.

5 This pattern is made from matchsticks.

    a How many matchsticks are there in the fifth pattern?

    b The number of matchsticks in the nth pattern is given by the expression $2n + 1$.
       How many matchsticks are there in:

       i the 10th pattern   **ii** the 15th pattern   **iii** the 20th pattern?

6 Look at these patterns, made from matchsticks.

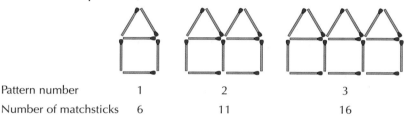

Pattern number	1	2	3
Number of matchsticks	6	11	16

    a How many matchsticks are there in the fourth pattern?

    b How many matchsticks are there in the fifth pattern?

    c The number of matchsticks in the nth pattern is given by the expression $5n + 1$.
       How many matchsticks are there in the:

       i 10th pattern   **ii** 20th pattern   **iii** 40th pattern?

 7 Write down any four consecutive terms from the Fibonacci sequence.

Multiply the first and the last terms together.

Multiply the middle two terms together.

What is the difference between these two numbers?

Try this rule for at least another three sets of four consecutive Fibonacci terms.

Write down what you notice.

Investigation
Pond borders

Square stone paving slabs each have an area of 1 m^2 and are used to put borders around a square pond.

Here is an example for a square pond measuring 3 m by 3 m.

16 slabs fit around the pond.

How many slabs would fit around a 6 m by 6 m pond?

To solve this problem you need to draw diagrams first for different sizes of square ponds.

1 On squared paper, copy the diagrams and fill in the two missing numbers.

  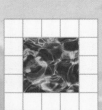 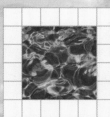

1 m by 1 m	2 m by 2 m	3m by 3m	4 m by 4 m
8 slabs	12 slabs	? slabs	? slabs

2 Copy and complete the table to show your results.

3 The number of slabs for different ponds form
a sequence.
Write down the term-to-term rule for the sequence.

4 Now write down the number of slabs that
would fit around a 6 m by 6 m pond.

Pond size	Number of slabs
1 m by 1 m	8
2 m by 2 m	12
3 m by 3 m	
4 m by 4 m	

5 Here are some rectangular ponds.

      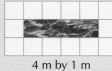   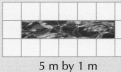

2 m by 1 m   3 m by 1 m   4 m by 1 m   5 m by 1 m

Work out how many 1 m^2 slabs fit around an 8 m by 1 m pond.

6

Area

This chapter is going to show you:

- how to use a formula to work out the area of a rectangle
- how to work out the area of a compound shape
- how to use a formula to work out the area of a triangle
- how to use a formula to work out the area of a parallelogram.

You should already know:

- how to work out the area of a rectangle using area = length × width
- that the units of area are cm² and m²
- how to substitute numbers into a formula.

About this chapter

The Forth Railway Bridge crosses the Firth of Forth in the east of Scotland. The structure uses triangles to give the bridge strength.

It used nearly 58 000 tonnes of metal and 18 122 m³ of granite and has a total length of 2528.7 metres.

We sometimes say a never-ending task is like 'painting the Forth Bridge'. This saying is based on the old – but false – belief that it takes so long to paint it that as soon as you finish one coat you have to start again.

A recent repainting involved applying 230 000 m² of paint, at a total cost of £130 million. This new coat of paint is expected to last at least 25 years, and possibly 40. Engineers had to work out the total surface area of the bridge to calculate how much paint was needed.

6.1 Area of a rectangle

Learning objective

- To use a formula to work out the area of a rectangle

Key word

formula

You already know that the area of the rectangle is the amount of space inside it.

Each square on this grid represents 1 cm^2.

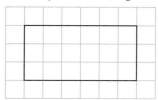

There are 6 squares on each row and there are 3 rows.

So the area of the rectangle is 18 cm^2.

You could just work out the area as $6 \times 3 = 18$ cm^2.

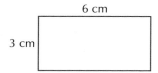

6 cm

3 cm

For any rectangle:

area = length × width

Remember that the metric units for area are usually the square centimetre (cm^2) or the square metre (m^2).

You can now use algebra to write the area as a **formula**.

When area = A, length = l and width = w, then $A = l \times w$ or $A = lw$.

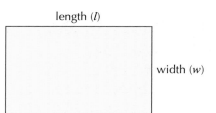

length (l)

width (w)

Example 1

Use the formula $A = lw$ to work out the area of each rectangle.

a
8 cm

3 cm

b
5 m

4 m

a $A = 8 \times 3 = 24$ cm^2

b $A = 5 \times 4 = 20$ m^2

Example 2

Use the formula $A = lw$ to work out the area of each square.

a 3 cm

3 cm

b 5 m

5 m

For a square the length and width are the same.

a $A = 3 \times 3 = 9 \text{ cm}^2$ **b** $A = 5 \times 5 = 25 \text{ m}^2$

Example 3

This rectangular garden has an area of 40 m².
Work out the width, w, of the garden.

$A = lw$, so $8w = 40$

Now divide both sides by 8 to calculate w.

So $w = 40 \div 8 = 5$ m.

8 m

w

Exercise 6A

1 Work out the area of each rectangle.

a 7 cm, 2 cm

b 12 cm, 8 cm

c 8 cm, 12.5 cm

  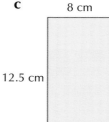

d 15 m, 9 m

e 3.5 m, 1 m

f 1.5 m, 5 m

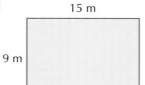 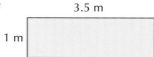 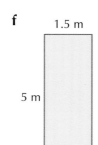

2 Work out the area of each square.

a 2 cm

b 4 cm

c 7 m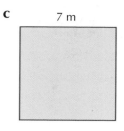

d 8.5 m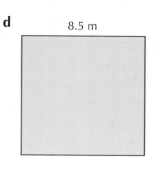

3 Work out the length, l, of each rectangle.

a
l
4 cm | Area = 28 cm²

b
l
7 cm | Area = 56 cm²

c
l
6 m | Area = 90 m²

d
l
9 m | Area = 144 m²

4 Work out the width, w, of each rectangle.

a
6 cm
w | Area = 24 cm²

b
5 cm
w | Area = 5 cm²

c
7 m
w | Area = 42 m²

d
30 m
w | Area = 360 m²

5 Copy and complete the table for rectangles **a** to **f**.

	Length	Width	Area
a	7 cm	5 cm	
b	10 cm	9 cm	
c		7 cm	63 cm²
d		8 cm	88 cm²
e	12 cm		60 cm²
f	25 cm		100 cm²

6 The diagram shows the measurements of a football pitch.

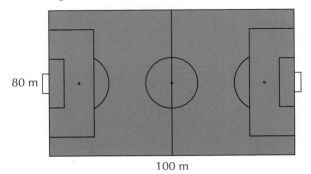

80 m

100 m

Work out the area of the pitch.

7 A rectangular carpet has an area of 75 m². The length of the carpet is 25 m.
Calculate the width of the carpet.

PS 8 Work out the perimeter of this square tile.

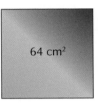

64 cm²

Problem solving: Finding lengths and widths

A A rectangular photograph has an area of 90 cm².
Its length is 1 cm more than its width.
Work out the length and width of the photograph.

B A rectangular playground has an area of 600 m².
Its length is 10 m more than its width.
Work out the length and width of the playground.

6.2 Areas of compound shapes

Learning objective

• To work out the area of a compound shape

Key word

compound shape

A **compound shape** is made from more than one shape. You can work out its area by splitting it into the shapes that make it up.

These two rectangles can be put together to make a compound shape.

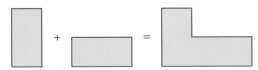

Example 4

The two rectangles can be put together to make a compound shape.

What is the area of the compound shape?

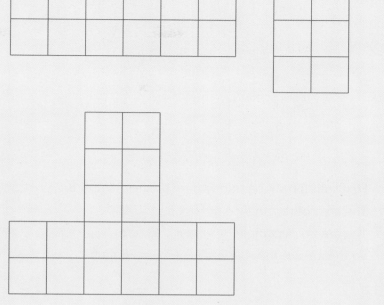

The first rectangle has an area of 12 cm^2.

The second rectangle has an area of 6 cm^2.

The area of the compound shape is 12 + 6 = 18 cm^2.

Example 5

Here are two rectangles A and B.

The two rectangles are put together to make a T-shape.

Work out the area of the T-shape.

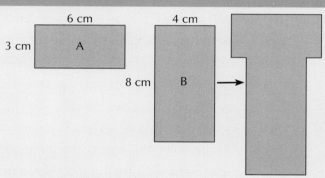

The T-shape can be treated as two rectangles.

The area of rectangle A = 6 × 3 = 18 cm^2.

The area of rectangle B = 8 × 4 = 32 cm^2.

So the area of the T-shape = 18 + 32 = 50 cm^2.

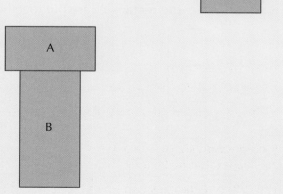

Example 6

Work out the area of this compound shape.

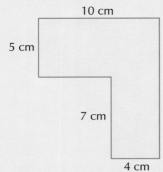

First split the shape into two rectangles, A and B.

The area of rectangle A = 10 × 5 = 50 cm².

The area of rectangle B = 7 × 4 = 28 cm².

So the area of the shape = 50 + 28 = 78 cm².

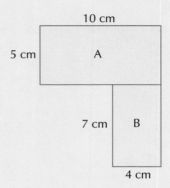

Exercise 6B

1 These three rectangles are made from centimetre squares.

a Write down the area of the three rectangles.

b The rectangles are put together to make compound shapes.

i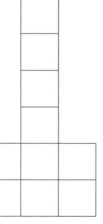

ii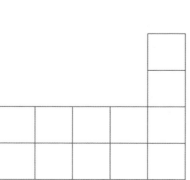

iii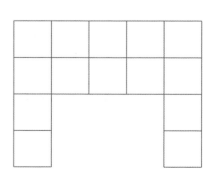

Write down the area of each compound shape.

2 Here are two rectangles A and B.

8 cm

5 cm   A

4 cm

12 cm   B

a The two rectangles are put together to make an L-shape.

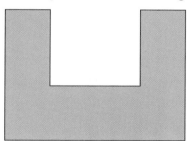

Work out the area of the L-shape.

b Rectangle A and two rectangle Bs are put together to make a U-shape.

Work out the area of the U-shape.

3 Here are three squares X, Y and Z.

They are put together to make a pyramid shape.

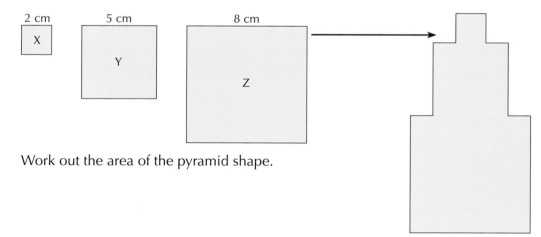

2 cm

X

5 cm

Y

8 cm

Z

Work out the area of the pyramid shape.

4 Work out the area of each compound shape.

a
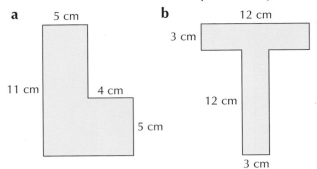

b

c
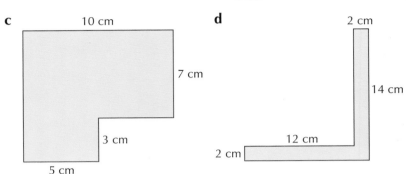

d

5 A path goes round a rectangular lawn.
The path is 1 m wide.

a Write down the area of the lawn.

b Work out the area of the path.

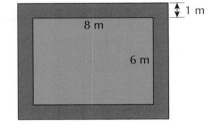

6 The diagram shows a compound shape.

a Work out the area of the shape.   **b** Work out the perimeter of the shape.

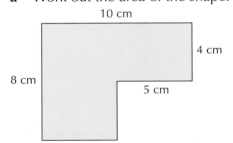

 Hint Work out the lengths of the two missing side first.

Reasoning: Area subtraction

Work out the area of the yellow shaded part of this shape.

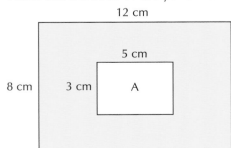

Area of complete rectangle = 12 × 8 = 96 cm^2

Area of rectangle A = 5 × 3 = 15 cm^2

So shaded area = 96 − 15 = 81 cm^2.

Work out the area of the shaded part for each shape.

A

10 cm

5 cm 6 cm 3 cm

B

5 cm

2 cm

9 cm 6 cm

C

8 cm

4 cm

8 cm 4 cm

6.3 Area of a triangle

Learning objective

- To use a formula to work out the area of a triangle

Here is a right-angled triangle. How do you work out its area?

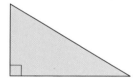

Two of the triangles fit together to make a rectangle.

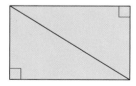

So to work out the area of the triangle, first work out the area of the rectangle and then divide the answer by 2.

It is usual to refer to the length of the triangle as its **base** and the width as its **height**.

The height is also referred to as the **perpendicular height**, because the base is perpendicular to the height.

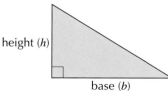

height (*h*)

base (*b*)

So to work out the area of the right-angled triangle, multiply the base by the height and then divide by 2.

$$\text{area} = \frac{\text{base} \times \text{height}}{2}$$

You can write this as a formula:

$$A = \frac{bh}{2}$$

This method can be used to work out the area of any triangle.

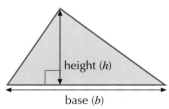

height (h)

base (b)

This diagram shows that the area of the triangle is half of the area of a rectangle that encloses the triangle.

area of triangle 1 = area of triangle 2 and

area of triangle 3 = area of triangle 4

You can find the area of the rectangle by multiplying the base by the height.

So, the area of the triangle is:

$$\text{Area} = \frac{\text{base} \times \text{height}}{2}$$

The formula for the area of a triangle is:

$$A = \frac{b \times h}{2} = \frac{bh}{2}$$

This is true for all triangles.

Example 7

Work out the area of this right-angled triangle.

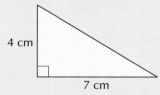

4 cm

7 cm

The formula for the area of the triangle is: $A = \frac{bh}{2}$

So $A = \frac{7 \times 4}{2} = \frac{28}{2} = 14 \text{ cm}^2$.

Example 8

Work out the area of this triangle.

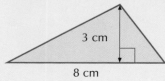

3 cm

8 cm

The formula for the area of the triangle is: $A = \frac{bh}{2}$

So $A = \frac{8 \times 3}{2} = \frac{24}{2} = 12 \text{ cm}^2$.

1 **a** Alex draws a rectangle on a centimetre square grid.

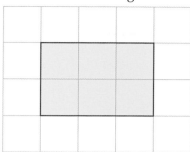

Write down the area of the rectangle.

b Alex then draws a right-angled triangle on the same grid.

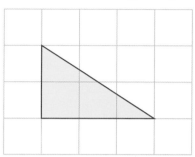

Write down the area of the triangle.

2 **a** Josh draws a rectangle on a centimetre square grid.

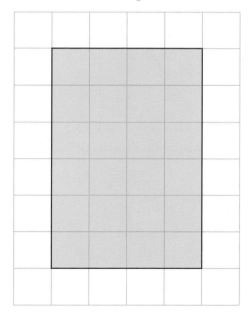

Write down the area of the rectangle.

b Josh then draws a right-angled triangle on the same grid.

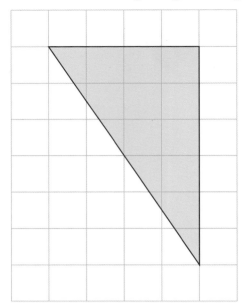

Write down the area of the triangle.

3 For each shape:
  i write down the area of the rectangle
  ii write down the area of the enclosed triangle.

a

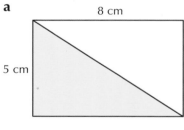

8 cm

5 cm

b

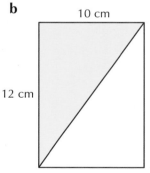

10 cm

12 cm

c

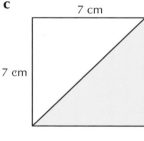

7 cm

7 cm

4 Copy and complete the calculations to work out the area of each right-angled triangle.

a

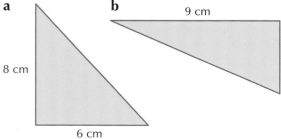

8 cm

6 cm

b

9 cm

4 cm

c

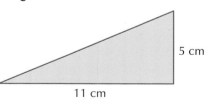

11 cm

5 cm

$A = \dfrac{bh}{2}$

So $A = \dfrac{\square \times \square}{2} = \dfrac{\square}{2} = \dots$ cm². So $A = \dfrac{\square \times \square}{2} = \dfrac{\square}{2} = \dots$ cm². So $A = \dfrac{\square \times \square}{2} = \dfrac{\square}{2} = \dots$ cm².

$A = \dfrac{bh}{2}$

$A = \dfrac{bh}{2}$

5 Work out the area of each triangle.

a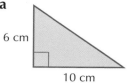

6 cm
10 cm

b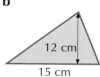

12 cm
15 cm

c

7 cm
6 cm

d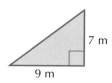

7 m
9 m

e

7 m
8 m

(MR) 6 This right-angled triangle has an area of 24 cm².

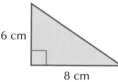

6 cm
8 cm

Find other right-angled triangles, with different measurements, that also have an area of 24 cm².

(PS) 7 Work out the area of each compound shape.

a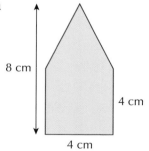

8 cm
4 cm
4 cm

b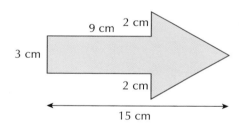

9 cm   2 cm
3 cm
2 cm
15 cm

Investigation: Compound triangles

This compound shape is made from four different right-angled triangles P, Q, R and S.

A Work out the area of the four right-angled triangles.

B Write down the area of the compound shape.

C The areas of the triangles making the shape form a sequence.

  a Write down this sequence.

  b What is the term-to-term rule for this sequence?

  c Write down the next number in this sequence.

D Another larger right-angled triangle is added to the shape, following the same pattern.

  What is the area of the new compound shape?

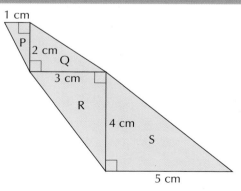

1 cm
P
2 cm
Q
3 cm
R
4 cm
S
5 cm

6.4 Area of a parallelogram

Learning objective

- To work out the area of a parallelogram

Key word

parallelogram

To work out the area of a **parallelogram**, you need to know the length of its base and its height (or perpendicular height).

These diagrams show that the parallelogram has the same area as a rectangle with the same base and height.

The parallelogram has a base, *b*, and a height, *h*.

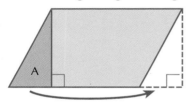

Move the right-angled triangle A to the other side of the parallelogram.

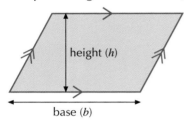

This now makes a rectangle with the same area as the parallelogram.

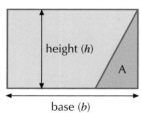

The parallelogram has the same area as the rectangle.

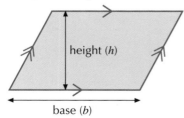 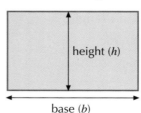

So the area of a parallelogram is:

base × height

The formula for the area of a parallelogram is:

$A = b \times h = bh$

Work out the area of this parallelogram.

$A = 6 \times 10 = 60 \text{ cm}^2$

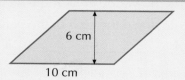

This parallelogram has an area of 56 cm².

Work out the height, h, of the parallelogram.

$A = 56 \text{ cm}^2$, so $h = 56 \div 8 = 7 \text{ cm}$

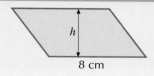

Exercise 6D

1   **a**   Sam draws a rectangle on a centimetre-square grid.

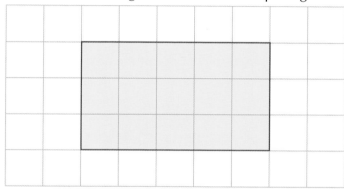

Write down the area of the rectangle.

    b   He then draws a parallelogram on the same grid.

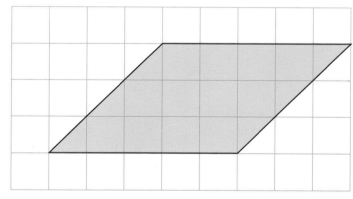

By counting squares, work out the area of the parallelogram.

2 For each pair of shapes:

   i write down the area of the rectangle

   ii write down the area of the parallelogram.

a i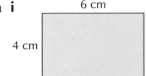
6 cm
4 cm

ii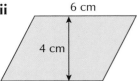
6 cm
4 cm

b i
9 cm
3 cm

ii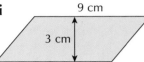
9 cm
3 cm

c i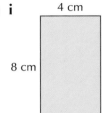
4 cm
8 cm

ii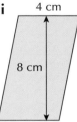
4 cm
8 cm

3 Copy and complete the calculations to work out the area of each parallelogram.

a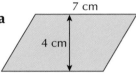
7 cm
4 cm

b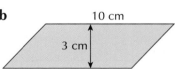
10 cm
3 cm

c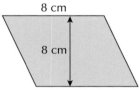
8 cm
8 cm

$A = bh$
So $A = ... \times ... = ...$ cm^2.

$A = bh$
So $A = ... \times ... = ...$ cm^2.

$A = bh$
So $A = ... \times ... = ...$ cm^2.

4 Copy and complete the calculations to work out the height, h, for each parallelogram.

a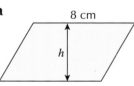
8 cm
h

b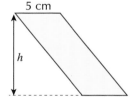
5 cm
h

c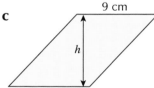
9 cm
h

$A = 32$ cm^2
So $h = ... \div ... = ...$ cm.

$A = 60$ cm^2
So $h = ... \div ... = ...$ cm.

$A = 63$ cm^2
So $h = ... \div ... = ...$ cm.

5 Copy and complete the calculations to work out the base, b, for each parallelogram.

a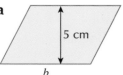
5 cm
b

b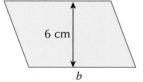
6 cm
b

c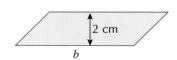
2 cm
b

$A = 35$ cm^2
So $b = ... \div ... = ...$ cm.

$A = 48$ cm^2
So $b = ... \div ... = ...$ cm.

$A = 9$ cm^2
So $b = ... \div ... = ...$ cm.

6 Work out the area of each parallelogram.

a
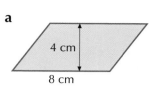
4 cm
8 cm

b
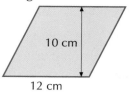
10 cm
12 cm

c
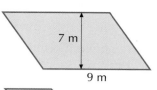
7 m
9 m

d
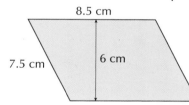
20 cm
5 cm

e
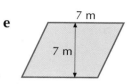
7 m
7 m

f
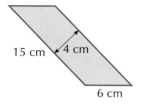
15 cm
4 cm
6 cm

(MR) 7 The area of this parallelogram is 20 cm².
Work out the height, h, of the parallelogram.

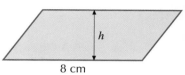
h
8 cm

(PS) 8 Work out: **a** the perimeter **b** the area of this parallelogram.

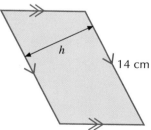
8.5 cm
7.5 cm
6 cm

Challenge: Problems to work out the height in parallelograms

A The two parallelograms shown each have an area of 56 cm².

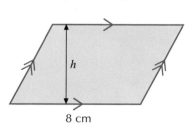
h
8 cm

h
14 cm

Work out the height, h, of each parallelogram.

B Work out the value of h in this diagram.

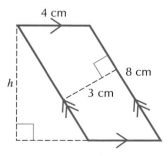
4 cm
8 cm
3 cm
h

Ready to progress?

Review questions

1 a On a centimetre-square grid, draw three different rectangles which each have an area of 12 cm².

   b Work out the perimeter of each of your rectangles.

2 Work out the area of each rectangle.

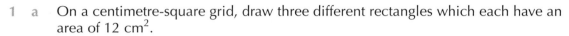

a 7 cm 3 cm

b 8 m 4 m

c 6.5 cm 10 cm

3 Here is a square tile.

Megan has a collection of these tiles and makes this cross shape using some of them.

Work out the area of the cross-shape.

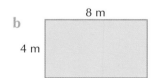

3 cm

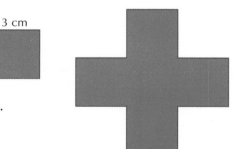

4 Copy and complete the following to work out the area of each right-angled triangle.

a
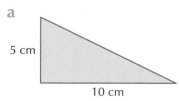
5 cm
10 cm

b
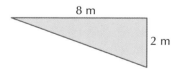
8 m
2 m

$A = \dfrac{bh}{2}$

So $A = \dfrac{\square \times \square}{2} = \dfrac{\square}{2} = \ldots$ cm².

$A = \dfrac{bh}{2}$

So $A = \dfrac{\square \times \square}{2} = \dfrac{\square}{2} = \ldots$ m².

5 Copy and complete the following to work out the area of each parallelogram.

a

$A = bh$
So $A = \ldots \times \ldots = \ldots$ cm^2.

b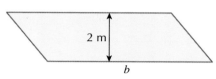

$A = bh$
So $A = \ldots \times \ldots = \ldots$ m^2.

6 Copy and complete the following to work out the base, b, for each parallelogram.

a

b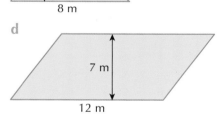

$A = 36$ cm^2
So $b = \ldots \div \ldots = \ldots$ cm.

$A = 15$ m^2
So $b = \ldots \div \ldots = \ldots$ m.

7 Work out the area of each shape.

a

9 cm

10 cm

b

7 m

8 m

c

8 cm

4 cm

d

7 m

12 m

8 Work out the area of this kite, made from two different isosceles triangles.

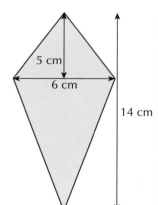

5 cm

6 cm

14 cm

Investigation
Pick's formula

The shapes below are drawn on a centimetre-squared grid of dots.

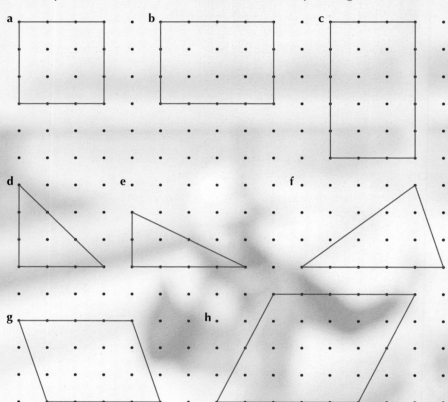

1 Copy and complete the table, filling in the numbers for each
   shape. The first row has been done for you.

Shape	Number of dots on perimeter of shape (P)	Number of dots inside shape (I)	Area of shape (A) (cm²)
a	12	4	9
b			
c			
d			
e			
f			
g			
h			

2 The formula that connects the area, A, to the number of dots on the perimeter, P and the number of dots inside, I is:

$$A = \frac{P}{2} + I - 1$$

Check that the formula works for the examples in question 1.

For example in **a**: $A = \frac{12}{2} + 4 - 1 = 6 + 4 - 1 = 9 \text{ cm}^2$.

This is known as Pick's formula, after Georg Alexander Pick, a famous Austrian mathematician. He was born in 1859 in Vienna and died in 1943 in Germany.

The formula is used to work out the areas of irregular polygons in surveying and map making.

3 Use Pick's formula to work out the area of the following shapes drawn on a centimetre-squared grid of dots..

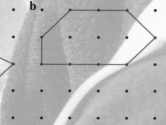 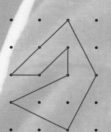 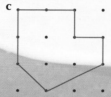

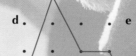 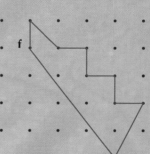

Record your results on a copy of this table.

Shape	Number of dots on perimeter of shape (P)	Number of dots inside shape (I)	Area of shape (A) (cm²)
a			
b			
c			
d			
e			
f			

7

Graphs

This chapter is going to show you:
- how to draw graphs of linear equations
- how to draw graphs of simple quadratic equations
- how to draw graphs to illustrate real-life situations.

You should already know:
- how to plot coordinates in all four quadrants
- how to calculate with negative numbers.

About this chapter

Equations are a powerful mathematical tool used in design, engineering and computer software development. The fact that these equations can create graphs allows designers, engineers and developers to model real life scenarios on computers. The shapes made by different equations allow them to be creative and to show movement on screen. To do this, the designers and engineers have to understand the nature of different equations and the effect of changing variables in them.

7.1 Rules with coordinates

Learning objective

• To recognise patterns within coordinates

Number puzzles and sequences can be made in coordinates as well as ordinary numbers.

Example 1

There is a dot in each of the rectangles in the diagram. The dots are in the corners that are defined by the coordinates (2, 1), (4, 1), (6, 1), (8, 1).

a The pattern of rectangles continues. Write down the coordinates of the next four corners with dots.

b Explain how you can tell that there will be no dot at the coordinate (35, 1).

a Look at the *x*-coordinates (the left-hand numbers). The numbers go up in the sequence 2, 4, 6, 8 (even numbers). So, the next four will be 10, 12, 14 and 16.

Look at the *y*-coordinates (the right-hand numbers). The numbers are all 1.

So, the next four coordinates will be (10, 1), (12, 1), (14, 1), (16, 1).

b Since 35 is not an even number, (35, 1) cannot be the coordinates of a corner with a dot.

Exercise 7A

(MR) (PS) 1 Joy has 20 rectangular tiles like the one shown.

She places all these tiles in a row.

She starts her row like this.

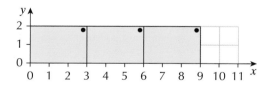

Joy writes down the coordinates of the corner of each tile which has a dot.

The coordinates of the corner with a dot in the first tile are (3, 2).

a Write down the coordinates of the next six corners that have a dot.

b Look at the numbers in the coordinates.

Describe two things you notice.

c Joy thinks that (41, 2) are the coordinates of one of the corners with a dot.

Explain how you know she is wrong.

d What are the coordinates of the corner with a dot in the 20th tile?

2 Joy now places her tiles in a pattern like this.

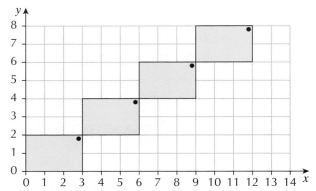

a Write down the coordinates of the first six corners which have a dot.

b Look at the numbers in the coordinates.
Describe two things you notice.

c Joy thinks that that (24, 16) are the coordinates of one of the corners with a dot.
Explain how you know she is right.

d What are the coordinates of the corner with the dot in the 20th tile?

3 Chris has 20 square tiles like this one.

2 cm

2 cm

He places all these tiles in a column.
He starts his pattern like this.

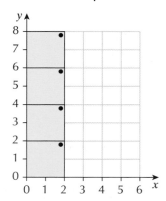

Chris writes down the coordinates of the corner on each tile which has a dot.
The coordinates of the first corner are (2, 2).

a Write down the coordinates of the next six corners that have a dot.

b Look at the numbers in the coordinates.
Describe two things you notice.

c Chris thinks that (2, 22) are the coordinates of one of the corners with a dot.
Explain how you know he is right.

d What are the coordinates of the corner with a dot in the 20th tile?

4 Andrew has 20 rectangular tiles like the one shown.

3 cm

2 cm

He places all these tiles in a pattern.

He starts his pattern like this.

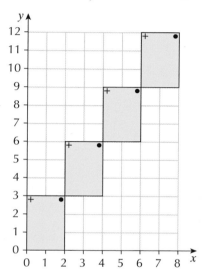

For each rectangular tile, Andrew writes down the coordinates of the corner which has a dot.

The coordinates of the first corner are (2, 3).

a Write down the coordinates of the next six corners which have a dot.

b Look at the numbers in the coordinates.

Describe two things you notice.

c Andrew thinks that (14, 21) are the coordinates of one of the corners which has a dot.

Explain how you know he is correct.

d Andrew's sister Sophia thinks that (39, 25) are the coordinates of one of the corners which has a dot.

Explain how you know she is not correct.

e What are the coordinates of the corner with a dot in the 20th tile?

  5 Sophia now looks at the tiles and writes down the coordinates of the corner which has a +.

 The coordinates of the first corner are (0, 3).

a Write down the coordinates of the next six corners which have a +.

b Look at the numbers in the coordinates. Describe two things you notice.

c Sophia thinks that (20, 33) are the coordinates of one of the corners which has a +.

Explain how you can tell she is correct.

d What are the coordinates of the corner with a + in the 20th tile?

(MR) **6** Tilly has 20 rectangular tiles like the one shown.

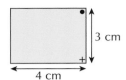

(PS)

She places all these tiles in a pattern.

She starts her pattern like this.

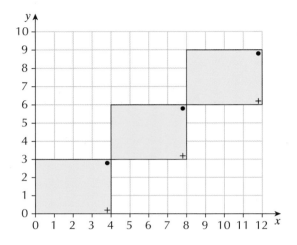

For each rectangular tile, Tilly writes down the coordinates of the corner which has a dot.

The coordinates of the first corner are (4, 3).

a Write down the coordinates of the next six corners that have a dot.

b Look at the numbers in the coordinates.

Describe two things you notice.

c Tilly thinks that (24, 10) are the coordinates of one of the corners which has a dot.

Explain why she is wrong.

d What are the coordinates of the corner with a dot in the 20th tile?

For each rectangular tile, Tilly now writes down the coordinates of the corner which has a +.

The coordinates of the first corner are (4, 0).

e Write down the coordinates of the next six corners that have a +.

f Look at the numbers in the coordinates. Describe two things you notice.

g Tilly thinks that (28, 16) are the coordinates of one of the corners which has a +.

Explain why she is wrong.

h What are the coordinates of the corner with a + in the 20th tile?

7.2 Graphs from rules

Learning objective

- To draw graphs of linear rules

There are different ways to write rules. For example, the rule: ⟶ | Add 3 | ⟶ can also be
written as $y = x + 3$, where the inputs are x and the outputs are y. This form is called a **linear equation**.

This second way of writing a rule makes it easier to draw a graph of the rule.

You can draw a graph of the equation by working out a set of coordinates.

The graph of every linear equation is a straight line.

Example 2

Draw a graph of the rule ⟶ | Multiply by 3 | ⟶ or $y = 3x$.

First, draw up a table of simple values for x.

x	0	1	2	3
$y = 3x$	0	3	6	9

So, the coordinate pairs are:

(0, 0) (1, 3) (2, 6) (3, 9)

Next, plot each point on a grid, join up all the points, and label the line.

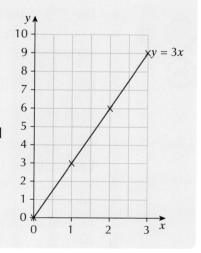

Notice that this straight line graph has hundreds of other coordinates, all of which obey the same rule, that is, $y = 3x$. Choose any points on the line that have not been plotted and you will see that this is true.

Exercise 7B

1 **a** Copy and complete the table for the rule ⟶ | Add 3 | ⟶ or $y = x + 3$.

x	0	1	2	3
$y = x + 3$	3			

b Draw a coordinate grid, numbering the x-axis from 0 to 3 and the y-axis from 0 to 7.

c Use values from the table to draw, on the grid, the graph of $y = x + 3$.

2 **a** Copy and complete the table below for the rule ⟶ | Multiply by 2 | ⟶ or $y = 2x$.

x	0	1	2	3
$y = 2x$	0			

b Draw a coordinate grid, numbering the x-axis from 0 to 3 and the y-axis from 0 to 7.

c Use values from the table to draw, on the grid, the graph of $y = 2x$.

3 **a** Copy and complete the table below for the function ⟶ | Subtract 1 | ⟶ or $y = x - 1$.

x	0	1	2	3	4
$y = x - 1$	−1				

b Draw a coordinate grid, numbering the x-axis from 0 to 4 and the y-axis from −1 to 4.

c Use values from the table to draw, on the grid, the graph of $y = x - 1$.

4 **a** Copy and complete the table for the rule ⟶ | Divide by 2 | ⟶ or $y = \frac{1}{2}x$.

x	0	1	2	3	4
$y = \frac{1}{2}x$	0	0.5			

b Draw a coordinate grid, numbering the x-axis from 0 to 4 and the y-axis from 0 to 4.

c Use values from the table to draw, on the grid, the graph of $y = \frac{1}{2}x$.

5 **a** Copy and complete the table for each equation.

x	0	1	2	3	4
$y = x + 5$				8	
$y = x + 3$			5		
$y = x + 1$	1	2			
$y = x - 1$	−1	0			
$y = x - 3$	−3		−1		

b Draw a coordinate grid, numbering the x-axis from 0 to 4 and the y-axis from −4 to 10.

c Use values from the table to draw, on the same grid, the graph for each equation.

d What two properties do you notice about each line?

e Use the properties you have noticed to draw the graphs of each equation.

　　i $y = x + 2.5$　　　**ii** $y = x - 1.5$

6 **a** Copy and complete the table for each equation.

x	0	1	2	3	4
$y = x$				3	
$y = 2x$			4		
$y = 3x$	0	3			
$y = 4x$					16
$y = 5x$			10		

b Draw a coordinate grid, numbering the x-axis from 0 to 4 and the y-axis from 0 to 20.

c Use values from the table to draw, on the same grid, the graph for each equation.

d What do you notice about each line?

e Use the properties you have noticed to draw the graphs of each equation.

   i $y = 1.5x$ **ii** $y = 3.5x$

 7 **a** Copy and complete the table below for the equations shown.

x	0	1	2	3	4
$y = 2x + 5$				11	
$y = 2x + 3$			7		
$y = 2x + 1$	1	3			
$y = 2x - 1$	−1	1			
$y = 2x - 3$	−3		1		

b Draw a coordinate grid, numbering the x-axis from 0 to 4 and the y-axis from −4 to 14.

c Use values from the table to draw, on the same grid, the graph for each equation.

d What two properties do you notice about each line?

e Use the properties you have noticed to draw the graph of each equation.

   i $y = 2x + 3.5$ **ii** $y = 2x - 1.5$

 8 **a** Copy and complete the table below for the equations shown.

x	0	1	2	3
$y = x + 1$				4
$y = 2x + 1$			5	
$y = 3x + 1$		4		
$y = 4x + 1$		5		

b Draw a coordinate grid, numbering the x-axis from 0 to 3 and the y-axis from 0 to 14.

c Use values from the table to draw, on the same grid, the graph for each equation.

d What properties do you notice about each line?

e Use the properties you have noticed to draw the graph of each equation.

   i $y = 1.5x + 1$ **ii** $y = 3.5x + 1$

7.3 Graphs from simple quadratic equations

Learning objective

- To recognise and draw the graph from a simple quadratic equation

Key word

quadratic

In a **quadratic** equation one of the variables is squared.

Here are some examples of simple quadratic equations.

- $y = x^2$
- $y = x^2 + 1$
- $y = 3x^2$
- $y = 4x^2 - 3$

You can follow the same technique of finding coordinates that fit the equation and plotting them on a graph. This time, however, the lines are not straight!

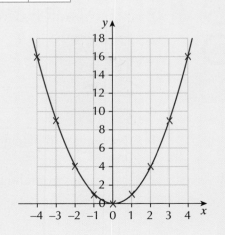
Notice the shape for a quadratic equation is a smooth curve. It is important to draw a quadratic graph very carefully, especially at the bottom of the graph where it needs to be a smooth curve.

Exercise 7C

1 a Copy and complete this table for the equation $y = x^2 + 1$.

x	−2	−1	0	1	2
x^2	4	1	0	1	4
$y = x^2 + 1$			1		

b Draw a coordinate grid, numbering the x-axis from −2 to 2 and the y-axis from 0 to 6.

c Use values from your table to draw, on your grid, the graph of $y = x^2 + 1$.

2 a Copy and complete this table for the equation $y = x^2 + 3$.

x	−2	−1	0	1	2
x^2					
$y = x^2 + 3$					

b Draw a coordinate grid, numbering the x-axis from −2 to 2 and the y-axis from 0 to 8.

c Use values from your table to draw, on your grid, the graph of $y = x^2 + 3$.

3 a Use your answers to questions 1 and 2 to help you copy and complete this table for the equations shown.

x	−2	−1	0	1	2
$y = x^2 + 2$	6	3			
$y = x^2 + 4$					
$y = x^2 + 5$					

b Draw a coordinate grid, numbering the x-axis from −2 to 2 and the y-axis from 0 to 10.

c Draw, on your grid, the graph for each equation in the table.

d What do you notice about each line?

e Use what you have noticed to draw the graphs of these two equations.

　i $y = x^2 + 2.5$　　**ii** $y = x^2 + 3.5$

4 a Copy and complete this table for the equation $y = 2x^2$.

x	−2	−1	0	1	2
x^2	4	1	0	1	4
$y = 2x^2$			0		

b Draw a coordinate grid, numbering the x-axis from −2 to 2 and the y-axis from 0 to 10.

c Use values from your table to draw, on your grid, the graph of $y = 2x^2$.

5 a Copy and complete this table for the equation $y = 3x^2$.

x	−2	−1	0	1	2
x^2					
$y = 3x^2$	12		0		

b Draw a coordinate grid, numbering the x-axis from −2 to 2 and the y-axis from 0 to 12.

c Use values from your table to draw, on your grid, the graph of $y = 3x^2$.

Challenge: Ends and midpoints

A   Copy this diagram.

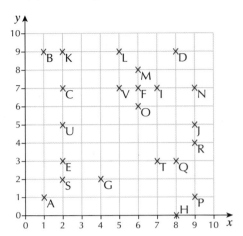

Join each pair of points to form eleven straight lines.

None of the lines should cross or touch another line.

i	A and B	**ii**	C and D	**iii**	E and F	**iv**	G and H
v	I and J	**vi**	K and L	**vii**	M and N	**viii**	O and P
ix	Q and R	**x**	S and T	**xi**	U and V		

B   Find the middle of each line.

Mark it with a dot.

C   Each line now has three coordinates, each marked with a dot.

Write each set of coordinates down like this.

For AB:

(1, 1)   (1, 9)

(1, 5)

The end points are on the top row, with the midpoint underneath, in the middle between them.

D   What do you notice about these sets of coordinates?

Try to write a rule about the coordinates of the end points and the midpoint.

E   Use your rule to predict what the coordinates of the middle point of these lines will be.

i	(1, 5) to (7, 5)	**ii**	(3, 4) to (6, 1)	**iii**	(–2, 1) to (–6, 3)
iv	(–3, –1) to (1, 2)	**v**	(–4, –2) to (0, 0)	**vi**	(1, –2) to (3, –1)
vii	(0, –5) to (3, –2)	**viii**	(–6, –4) to (–1, –4)	**ix**	(–5, –7) to (–1, –5)
x	(1, 7) to (7, –4)				

F   Draw the lines and check.

Were your predictions correct?

Distance–time graphs

l and to draw distance–time graphs

Key word

distance–time graph

e–time graph gives information about how someone or something has travelled. Therefore,
also called travel graphs.

–time graphs are often used to describe journeys.

le 4

et off from home at 8:00 am to pick up a parcel from the post office.

am he arrived at the post office, having walked 4 km.

ted 15 minutes, got his parcel and then walked back home in 20 minutes.

v a distance–time graph of Geza's journey.

your graph to estimate how far from home he was at:

0 am   **ii**   9:00 am.

e key coordinates (time and distance from home) are:

starting out (8:00 am, 0 km)

arrival at the post office (8:30 am, 4 km)

leaving the post office (8:45 am, 4 km)

return home (9:05 am, 0 km).

ing the information above, plot the points and draw the graph.

e axes should be labelled clearly and accurately, with precisely placed divisions.

s important to do this when you are drawing graphs.

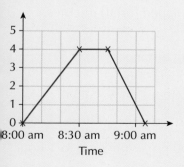

At 8:10 am, Geza was 1.3 km from home.

At 9:00 am, Geza was 1 km from home.

Exercise 7E

1

a Draw a coordinate grid with the following scales:

- horizontal axis: time, from 10:00 am to 11:30 am, with 1 cm to 10 minutes
- vertical axis: distance from home, from 0 to 8 km, with 1 cm to 1 km.

b Draw on your grid the travel graph which shows the following journey.

Gemma left home at 10:00 am.

She cycled 3 km to a friend's house, arriving there at 10:15 am.

The friend was not at home.

So Gemma cycled another 4 km in the same direction to another friend's arriving there at 10:35 am. This friend was in.

Gemma stayed there for 15 minutes.

Then she cycled back home, arriving at 11:30 am.

c Approximately how far from home was Gemma at the following times?

i 10:20 am   **ii** 11:15 am

d At what times, approximately, was Gemma 6 km away from home?

 2

a Draw a coordinate grid with the following scales:

- horizontal axis: time, from 8:00 am to 12 noon, with 1 cm to 30 minut
- vertical axis: distance, from 0 to 20 km, with 1 cm to 2 km.

b Draw on your grid the travel graph which shows the following journey.

Anne started her sponsored run at 8:00 am.

She ran the first 5 km in 30 minutes.

She ran the next 5 km in 45 minutes, then stopped to rest for 15 minutes.

She ran the next 5 km in 45 minutes, then stopped for 30 minutes.

She ran the last 5 km in 60 minutes.

c At about what time had she run the following distances?

i 3 km   **ii** 8 km   **iii** 17 km

d How far had she run after two hours?

 3

a Draw a grid with the following scales:

- horizontal axis: time, from 7:00 am to 4:00 pm, with 2 cm to 1 hour
- vertical axis: distance from Sheffield, from 0 to 600 km, with 1 cm to 5

b Draw on your grid the distance–time graph which shows the following jo

A bus set off from Sheffield at 7:00 am, travelling to Cornwall.

At 9:00 am the bus had travelled 150 km. It then stopped for 30 minutes.

At 12:00 noon the bus had travelled a further 200 km. It again stopped fo 30 minutes.

The bus arrived at its destination at 4:00 pm, after a journey of 600 km.

c Approximately how far from Sheffield was the bus at these times?

    i 8:00 am   **ii** 11:30 am   **iii** 2:00 pm

d Approximately, at what time was the bus 50 km away from its destination?

 4 **a** Draw a grid with the following scales:

- horizontal axis: time, from 2:00 pm to 5:00 pm, with 1 cm to 30 minutes
- vertical axis: distance from home, from 0 to 60 km, with 1 cm to 10 km.

b Draw on your grid the travel graph which shows the following journey.

Assume the seagull is always flying in the same direction out to sea.

A seagull left its nest at 2:00 pm, flying out to sea.

After 30 minutes, the bird had flown 10 km.

It then stopped on the top of a lighthouse for 30 minutes.

The bird then kept flying out to sea and at 4:00 pm landed on an anchored boat mast, 25 km from its nest.

The wind picked up and the bird flew again out to sea, stopping at 5:00 pm, 60 km from its nest.

c How far was the bird from its nest at the following times?

    i 3:30 pm   **ii** 4:45 pm

d At what time was the seagull 50 km away from its nest?

Problem solving: Missing friends

At 10:00 am, Joy set off from her home to walk towards Vicky's home, 500 metres away. At the same time, Vicky set off from her home to walk towards Joy's home.

Both girls were walking with their eyes down and did not see each other as they passed. Joy arrived at Vicky's house at ten minutes past ten. Vicky arrived at Joy's house at six minutes past ten.

A At what time did they pass each other?

B For how long were they within 100 metres of each other?

Ready to progress?

I can spot patterns in coordinate diagrams.
I can complete a table of values for a simple linear relationship and use this to draw a graph of the relationship.
I can draw and interpret distance–time graphs that describe real-life situations.

I can complete a table of values for a simple quadratic equation and use this to draw a graph of the equation.

Review questions

1 The total cumulative worldwide sales of a certain brand of tablet computer are shown in the graph.

Use the graph to find the missing numbers from the sentences below.

In November 2010, there had been about...million tablets sold.

Two years later, the sales of tablets were about...million.

From February 2011 to February 2014, the number of tablets sold had increased by about...million.

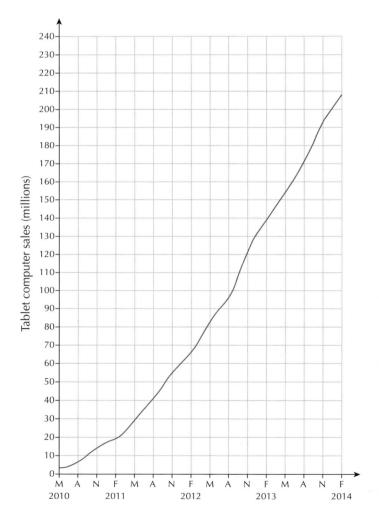

MR **2** Eve has 20 rectangular tiles like the one shown.

PS She places all these tiles in a diagonal pattern.

She starts her pattern as shown.

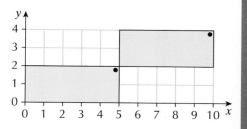

For each tile, Eve writes down the coordinates of the corner that has a dot.

The coordinates of the first corner are (5, 2).

a Eve thinks that (44, 16) are the coordinates of one of the corners with a dot.

Explain how you know she is wrong.

b What are the coordinates of the corner with a dot in the 20th tile?

3 **a** Draw a grid with the following scales.

- horizontal axis: time, from 11:00 am to 3:00 pm, with 1 cm to 30 minutes
- vertical axis: distance from home, from 0 to 40 km, with 1 cm to 5 km.

b Draw on your grid the travel graph which shows the following journey.

Brant left home at 11:00 am.

At 11:30 am he had cycled 5 km, when he stopped to have a drink.

He stopped for 15 minutes.

Brant then cycled another 15 km in the same direction into the Peak District, stopping for his lunch at 12:30 pm.

After a 30 minute lunch, he set off again cycling for another 15 km.

At 2:00 pm he found he had a flat tyre.

He managed to get a lift from a friend and got home at 3 pm.

c Approximately how far from home was Brant at the following times?

i 12 noon   **ii** 1:30 pm

d At what times, approximately, was Brant 5 km away from home?

4 **a** Copy and complete this table for the equation $y = x + 4$.

x	0	1	2	3	4
$y = x + 4$					

b Draw a coordinate grid, numbering the x-axis from 0 to 4 and the y-axis from 0 to 10.

c Use values from your table to draw, on your grid, the graph of $y = x + 4$.

5 **a** Copy and complete this table for the equation $y = 5x^2$.

x	−2	−1	0	1	2
x^2					
$y = 5x^2$					

b Draw a coordinate grid, numbering the x-axis from −2 to 2 and the y-axis from 0 to 25.

c Use values from your table to draw, on your grid, the graph of $y = 5x^2$.

Problem solving

The M60

The M60 motorway is a 36-mile long motorway that circles Manchester.

It was named the M60 in 2000 after joining parts of the M62 and the M63 with other main motorways around Manchester.

There are currently 27 junctions around the M60. Junction 3 is one of the busiest as it is the link to and from Manchester Airport.

1 How long is the M60?

2 How old is the M60 now?

3 The legal speed limit on the motorway is 70 mph. How long would it take to drive around the M60 at the legal speed limit? Assume that there are no delays.
Give your answer in minutes.

$$\text{time} = \frac{\text{distance}}{\text{speed}}$$

4 Someone drives all the way round the M60 in thirty minutes. What would be their average speed?

$$\text{average speed} = \frac{\text{distance}}{\text{time}}$$

5 Jim took 15 minutes to travel from junction 4 to junction 17, a distance of 18 miles.
Did he exceed the speed limit? Explain how you know.

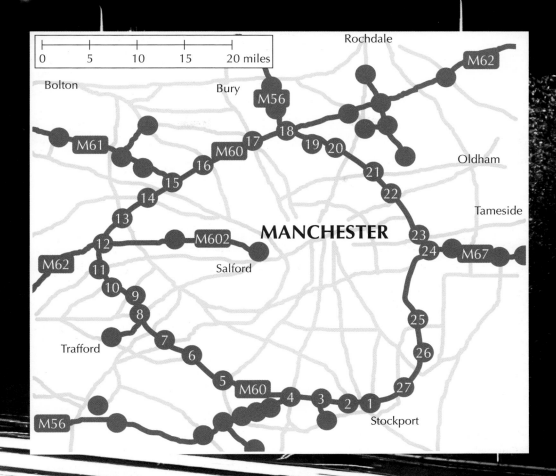

6 Brian left home to go to work in Salford at 6 am.
He drove the first 25 miles in 30 minutes, was held up at traffic lights
for ten minutes before slowly travelling the next 4 miles, through
Glossop, in 20 minutes. He finally got onto the M67 at 7 am. He
travelled the next 6 miles in 10 minutes before getting onto the M60 at
junction 24. The M60 was busy and he covered the 8 miles to junction
9 in thirty minutes. He did the final 3 miles to work in ten minutes.

   a Draw a distance–time graph to represent Brian's journey to work.

   b What time did he get to work?

   c How far did he travel from home to work?

   d How many junctions did he go through on the M60?

8

Simplifying numbers

This chapter is going to show you:

- how to multiply and divide by 10, 100 and 1000
- how to round numbers to one significant figure
- how to estimate answers to problems
- how to solve problems involving decimals.

You should already know:

- how to multiply and divide by 10
- how to round to one decimal place
- how to use standard column methods for the four operations.

About this chapter

Our nearest star, which is called Proxima Centauri, is an amazing 40 653 234 200 000 kilometres from Earth. An atom is 0.000 000 0001 metres wide.

When dealing with very large and very tiny numbers, it is easier to round them and present them as small numbers multiplied or divided by powers of 10. This is called standard form. It is a powerful tool that is widely used in science, especially by scientists such as astronomers who are dealing with massive spaces, or cell biologists, who are dealing with tiny ones. In this chapter, you will learn how to work with powers of 10.

8.1 Powers of 10

Learning objectives

- To multiply and divide by 100 and 1000
- To round numbers to one decimal place

You should remember:

- To multiply by 10, move all the digits 1 place to the left.
- To multiply by 100, move all the digits 2 place to the left.
- To multiply by 1000, move all the digits 3 place to the left.
- To divide by 10, move all the digits 1 place to the right.
- To divide by 100, move all the digits 2 place to the right.
- To divide by 1000, move all the digits 3 place to the right.

Notice you move the digits the same number of places as there are zeros after the 1.

Example 1

Round each number to one decimal place (1 dp).

a 9.35   **b** 4.323   **c** 5.99

To round to one decimal place, you need to look at the value of the digit in the second decimal place.

When the value of the digit is 5 or more then round up

When the value of the digit is less than 5 then round down.

a 9.35 is 9.4 to 1 dp   The value of the digit in the second decimal place is 5 so round up.

b 4.323 is 4.3 to 1 dp   The value of the digit in the second decimal place is 2 so round down.

c 5.99 is 6.0 to 1 dp   The value of the digit in the second decimal place is larger than 5 so round up.

Exercise 8A

1   Multiply each of these numbers by: **i** 10   **ii** 100   **iii** 1000.

     a 5.3   **b** 0.79   **c** 24   **d** 5.063   **e** 0.003

2   Divide each of these numbers by: **i** 10   **ii** 100.

     a 83   **b** 4.1   **c** 457   **d** 6.04   **e** 34 781

3   Write down the answers to each of these.

     a 3.1×10 **b** 6.78×100 **c** 0.56×1000 **d** $34 \div 100$

     e $823 \div 100$ **f** $9.06 \div 10$ **g** 57.89×100 **h** $57.89 \div 100$

     i 0.038×1000 **j** $0.038 \div 10$ **k** 0.05×1000 **l** $543 \div 100$

4   Round each number to one decimal place.
    a   4.722   b   3.097   c   2.634   d   1.932   e   0.784
    f   0.992   g   3.999   h   2.604   i   3.185   j   3.475

5   Round each number to: i   the nearest whole number   ii   one decimal place.
    a   4.72   b   3.07   c   2.634   d   1.932   e   0.78   f   0.92
    g   3.92   h   2.64   i   3.18   j   3.475   k   1.45   l   1.863

6   Multiply each number by: i   10   ii   100.
    Round each answer to one decimal place.
    a   0.4717   b   2.6345   c   0.0482

7   Divide each number by: i   10   ii   100.
    Round each answer to one decimal place.
    a   12.34   b   136.71   c   10.05

8   Explain why 2.99, becomes 3.0 when rounded off to 1 decimal place.

(PS) 9   State which of the following calculations gives:
    a   the largest answer   b   the smallest answer.
        A   3.142 × 10   B   0.0478 × 100   C   2417 ÷ 100

(PS) 10   Write down a number that will round to 5.5 to one decimal place when divided
    by 1000.

Challenge: Rounding to 2 dp

Sometimes you need to round numbers to two decimal places. An obvious case is when you work with money. Here are two examples of rounding to two decimal places (2 dp):

- 5.642 becomes 5.64 to 2 dp
- 8.776 becomes 8.78 to 2 dp.

Round each number to two decimal places.

A   4.722   B   3.097   C   2.634   D   1.932   E   0.784
F   0.992   G   3.999   H   2.604   I   3.185   J   3.475

8.2 Large numbers and rounding

Learning objective

- To round large numbers

Key words	
approximate	estimate
round	

When you are discussing large quantities, you often only need to use an **approximate** number. To work this out, you **round** the number up or down to the nearest suitable figure.

You can also use rounded numbers to **estimate** the answers to questions.

bar chart shows the annual profits for a large
pany over the previous five years. Estimate the
t for each year.

company chairman says: 'Profit in 2013
nearly 50 million pounds.' Is the chairman
ect?

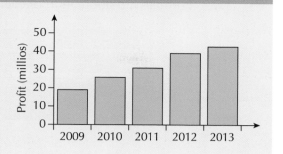

2009 the profit was about 19 million pounds.

2010 it was about 25 million pounds.

2011 it was about 31 million pounds.

2012 it was about 39 million pounds.

2013 it was about 43 million pounds.

e chairman is wrong, as in 2013 the profit is closer to 40 million pounds.

se 8B

Write each number in words.

a 4561	**b** 22 108	**c** 60 092	**d** 306 708
e 213 045	**f** 3 452 763	**g** 2 047 809	**h** 12 008 907

Write each number in figures.

a Four million, forty-three thousand, two hundred and seven.

b Nineteen million, five hundred and two thousand and thirty-seven.

c One million three hundred and two thousand and six.

Write down the answer to each problem, using figures.

a Seven thousand and two added to fifty thousand and eighty-three.

b Thirty-four thousand and sixty-one added to eighteen thousand, three hundred and two.

c Nine hundred and five thousand, seven hundred and thirteen added to one million.

Write down the answer to each problem, using words.

a One million, subtract two thousand and ninety-eight.

b Thirty-eight thousand, subtract twenty-three thousand and twelve.

c Three hundred and seven thousand, minus thirteen thousand and eighty.

5 The bar chart shows the population of some countries in the European Union
Estimate the population of each country.

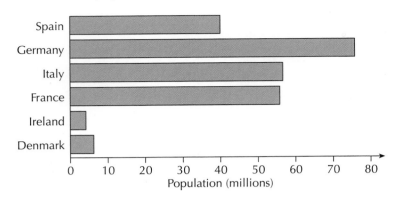

Population (millions)

6 Round each number to the nearest ten thousand.

a 67 821	**b** 42 321	**c** 546 231	**d** 782 700
e 603 000	**f** 256 403	**g** 3 547 812	**h** 9 722 1(
i 3 045 509	**j** 1 099 472	**k** 12 324 877	**l** 15 698 !

7 Round each number to the nearest hundred thousand.

a 256 000	**b** 748 989	**c** 521 200	**d** 591 333
e 4 678 923	**f** 8 631 095	**g** 4 153 410	**h** 5 030 8·
i 26 787 898	**j** 19 971 455	**k** 33 092 624	**l** 38 449]

8 Round each number to the nearest million.

a 2 436 701	**b** 6 833 217	**c** 2 531 298	**d** 4 198 0(
e 37 496 755	**f** 65 339 854	**g** 28 751 278	**h** 49 836]

(MR) **9** There are 2 452 800 people out of work. The government says: 'Unemploym
just over two million.' The opposition says: 'Unemployment is still nearly thre
million.' Who is correct and why?

(PS) **10** Look at the conversation.

What is the answer to the question?

8.3 Significant figures

Learning objective

* To round to one significant figure

Key words

significant

significant figure

As well as rounding to the nearest 10, 100, 1000, you can also use **significant figures** to round numbers. For example, when you say: 'It's about 2 miles to school,' or: 'It took me 30 minutes to do my homework last night,' you are rounding to one **significant** figure (1 sf).

In any number:

* the digit with the highest place value is the most significant figure.
* the digit with the next highest place value is the next most significant, and so on.

To round a number to one significant figure, you round the digit with the highest place value. For example, look at the number 382 649.

* The figure 3 has the highest place value, so it is the most significant. Rounded to one significant figure, 382 649 would be 400 000.

The method for rounding a number to significant figures is similar to the method for rounding to the nearest 10, 100, 1000 and so on.

* If the digit to the right of the significant number is 0, 1, 2, 3 or 4, then replace it by zero but leave the figure on the left unchanged.
* If this digit is 5, 6, 7, 8 or 9, then replace it by zero and add 1 to the figure before it, to the left.

A zero may or may not count as a significant figure, depending on its position.

When a zero is at the end of a whole number, it is not significant.

For example:

* 600 has one significant figure
* 720 has two significant figures.

When the number is less than 1, you start to count significant figures at the first non-zero digit. Any zeros that follow that digit count as significant.

For example:

* 0.007 has one significant figure
* 0.60 has two significant figures.

When zeros come between other digits, you count them as significant.

For example:

* 6.07 has three significant figures.

Example 3

Round:

a 34.87 to one significant figure (1 sf)   **b** 19 342 to one significant figure (1 sf).

   a To round 34.87 to 1 sf

      The most significant digit is 3.

      The figure to the right of it is 4.

      The value of this is less than 5, so there is no rounding up. The 3 stays the same.

      Put in a zero to preserve the place value of the 3.

      Hence $34.87 \approx 30$ (1 sf).

   b To round 19 342 to 2 sf

      The most significant digit is 1.

      The next figure to the right is 9.

      The value of this is greater than 5, so round up and add one to the 1, which makes 2.

      Replace the 9 and the other three figures with zeros to maintain the place value of the 2.

      Hence $19\,342 \approx 20\,000$ (1 sf).

Exercise 8C

1   State the number of significant figures in each number.
The first one has been done for you.

   a   1.3   **b**   325   **c**   5.24   **d**   0.509   **e**   8 million

> 1.3 has two significant figures

2   Round each number to one significant figure.
The first one has been done for you.

   a   327   **b**   3760   **c**   60.8   **d**   0.9137

> 3 is the most significant figure. The next figure to the right is 2, so round down.
> 327 rounded to one significant figure is 300.

   e   0.0853   **f**   257   **g**   68.9   **h**   3650
   i   0.7396   **j**   9.52   **k**   583.2   **l**   0.084

3   Calculate, giving your answers to one significant figure.

   a $53 + 18$ **b** $49 + 85$ **c** $95 + 123$ **d** $752 + 69$
   e $56 - 17$ **f** $735 - 108$ **g** $353 - 101$ **h** $649 - 214$
   i $488 + 127$ **j** $94 - 17$ **k** $869 + 315$ **l** $705 - 57$

4   Calculate, giving your answers to one significant figure.

   a 21×5 **b** 92×7 **c** 19×11 **d** 58×2
   e 94×3 **f** 25×6 **g** 26×4 **h** 109×8
   i 82×5 **j** 56×12 **k** 314×3 **l** 254×4

5 Calculate, giving your answers to one significant figure.

a 325 ÷ 5	**b** 755 ÷ 5	**c** 134 ÷ 2	**d** 171 ÷ 3
e 324 ÷ 4	**f** 368 ÷ 4	**g** 635 ÷ 5	**h** 754 ÷ 2
i 837 ÷ 10	**j** 10 097 ÷ 100	**k** 19.07 ÷ 10	**l** 23.1 ÷ 100

6 Use a calculator to work out each division. Round the answers to one significant figure.

a 1 ÷ 3	**b** 1 ÷ 7	**c** 1 ÷ 13	**d** 1 ÷ 11
e 11 ÷ 7	**f** 12 ÷ 13	**g** 17 ÷ 7	**h** 21 ÷ 11
i 51 ÷ 13	**j** 113 ÷ 37	**k** 187 ÷ 32	**l** 111 ÷ 33

(PS) 7 Pete painted his sitting room walls. He applied several coats of paint.

Altogether he used 7 litres of paint to cover an area of 75 m^2.

Work out the area, in square metres (m^2), that was covered by each litre. Give your answer to one significant figure.

(PS) 8 The population of a small country is 647 100. Its total land area is 5 600 000 m^2.

Work out, correct to one significant figure, how many square metres (m^2) there are to each person in this country.

9 Find the monthly pay, correct to one significant figure, of each person.

Name	Annual salary (£)
Peter	25 750
Paula	45 150
Reda	75 590
Arron	56 470
Melody	115 800

(PS) 10 Giles was planning to knock down a small wall. He wanted to know approximately how many bricks there were in the wall.

He counted about 215 bricks in one row.

He could see there were 9 rows of bricks in the wall.

How many bricks would he expect to get after knocking the wall down?

Give your answer correct to one significant figure.

Investigation: Patterns in calculations

A Use a calculator to work out the answer to each division.

Write down the calculator answer correct to one significant figure.

```
1 ÷ 9 =
1 ÷ 99 =
1 ÷ 999 =
1 ÷ 9999 =
1 ÷ 99999 =
1 ÷ 999999 =
```

B Explain any patterns that you notice.

C Write down the answer to 1 ÷ 99 999 999 correct to one significant figure.

D Write down the answer to 1 ÷ 999 999 999 correct to one significant figure.

E Investigate what happens when you use 6s instead of 9s.

8.4 Estimating answers

Learning objective

Key word

estimate

• To use rounding to estimate rough answers to calculations

You should know if the answer to a calculation is about the right size or not. Here are two ways of checking answers.

• In a multiplication, you can check that the final digit is correct by multiplying the last digit of each number together.

• Round off each number to one significant figure and do a mental calculation to see if the answer is about the right size. This is known as an **estimate** of the answer.

Example 4

Explain why these calculations must be wrong.

a $23 \times 45 = 1053$ **b** $19 \times 59 = 121$

a Multiply the last digit of each number together.

$3 \times 5 = 15$

The last digit should be 5, that is

$23 \times 45 = \ldots 5$

so the answer must be wrong.

b Round each number to one significant figure.

The answer is roughly $20 \times 60 = 1200$.

So the answer must be wrong.

Example 5

Estimate the answers to each calculation.

a $\dfrac{21.3 + 48.7}{6.5}$ **b** 31.2×48.5 **c** $359 \div 42$

a Round each number to one significant figure. The calculation is now:

$\dfrac{20 + 50}{7} = \dfrac{70}{7} = 10$

So an estimate of the answer is 10.

b Round each number to one significant figure. The calculation is now:

$30 \times 50 = 3 \times 5 \times 100 = 1500$

So an estimate of the answer is 1500.

c Round each number to the nearest 10. The calculation is now:

$360 \div 40 = 36 \div 4 = 9$

So an estimate of the answer is 9.

Exercise 8D

1 Explain why these calculations must be wrong.

  a $23 \times 48 = 1140$ **b** $61 \times 53 = 323$ **c** $36 \times 44 = 846$

  d $\dfrac{24.3 + 72.8}{8.9} = 20.07$ **e** $360 \div 8.9 = 54.4$ **f** $354 - 37 = 291$

2 Estimate the value the arrow is pointing at in each of these.

  a   **b**

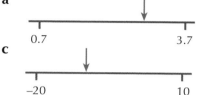

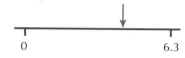

  c

3 Estimate the answer to each calculation, correct to one significant figure.

  a 27×3.9 **b** 23.1×1.8 **c** $812 \div 39$ **d** $\dfrac{26.8 + 34.3}{18.7}$

  e 312×312 **f** $46.2 \div 8.1$ **g** $\dfrac{46.8 - 8.3}{19.3 - 8.8}$ **h** $\dfrac{27.8 \times 5.3}{58 - 17.9}$

(MR) 4 Rafael had £5. He bought a magazine costing £2.15, some chocolate costing 87p, and a drink costing £1.49. Without adding up the numbers, how could Rafael be sure he had enough pay at the till?

(MR) 5 Hanna bought 5 cans of orangino at 37p per can. The shopkeeper asked her for £1.87. Without working out the correct answer, explain how Hanna could tell this was wrong.

(MR) 6 Megan needs five pens. A pen costs 37p. Will £2 be enough to pay for them? Explain your answer clearly.

(MR) 7 Tobiasz bought a train ticket for £4.45, paid with a £10 note and looked at the £5.45 change he had just been given. How did he know it was wrong?

(MR) 8 Laura bought seven ice creams for herself and her friends, at a cost of £1.15 each. She looked at the £10 note in her hand and hoped she had enough.

  How could she estimate if she had enough money to pay for them all?

(MR) 9 Barney had a £20 note in his pocket and went to the shops. He saw a ball at £6.75, a bat at £8.55 and a £5.20 set of stumps.

  Explain how Barney could estimate if he had enough money to buy them all.

(MR) 10 In a shop, Mazie bought a comic that cost 83p and a model that cost £1.57.

  When Mazie saw the till showing £84.57, she was shocked.

  Explain what had happened for the till to show that figure.

Challenge: Estimating squares

A Without working out areas or counting squares, explain why the area of the square shown must be between 36 and 64 grid squares.

B Now calculate the area of the square.

C Using an 8 by 8 grid, draw a square with an area of exactly 50 grid squares.

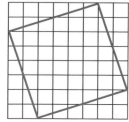

8.5 Problem solving with decimals

Learning objective

• To solve problems with decimal numbers

You can solve problems involving decimals.

Example 6

In a science lesson, Jason adds 0.15 kg of water and 0.04 kg of salt to a beaker that has a mass of 0.09 kg. He then pours out 0.12 kg of the mixture.

a What is the total mass of the beaker and mixture remaining?

b There are 28 pupils in the class and each pupil carries out the experiment. What is the total mass of salt needed for the class?

  a This has to be set up as an addition and subtraction problem. That is:

    $0.15 + 0.04 + 0.09 - 0.12$

    The problem has to be done in two stages.

$$
\begin{array}{ll}
0.15 & 0.28 \\
0.04 & -\,0.12 \\
+\,0.09 & \overline{0.16} \\
\overline{0.28} & \\
\end{array}
$$

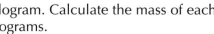

    So the final mass is 0.16 kg or 160 grams.

  b Total amount of salt needed = amount of salt needed for each pupil × number of pupils

    $= 0.04 \times 28$

    $= 1.12$ kg

Exercise 8E

1   Work these out.

  a $8.3 + 4.6$ **b** $8.3 - 4.9$ **c** $5.1 + 2.6 + 1.4$

  d $4.5 - 1.2 - 2.3$ **e** $8.2 - 2.9 - 2.7$ **f** 4.5 cm $+ 2.1$ cm $+ 8.6$ cm

2   Work these out.

  a $4.32 + 65.09 + 172.3$ **b** $8.7 + 9 + 14.02 + 1.03$ **c** $11.42 + 15.72 - 12.98$

  d $42.7 + 67.3 - 35.27$ **e** $19.87 + 2.8 - 13.46$ **f** $21.35 + 6.72 - 12.36 - 5.71$

3   There are 1000 metres in a kilometre. Work these out. Work in kilometres.

  a 7.45 km $+ 843$ m $+ 168$ m   **b** 3.896 km $+ 723$ m $+ 292$ m

  c 8.76 km $+ 463$ m $- 892$ m   **d** 16 km $- 435$ m $- 689$ m

  e 7.8 km $+ 5.043$ km $- 989$ m   **f** 6.85 km $+ 43$ m $- 89$ m

> **Hint**   Convert quantities so that they are all in the same units before you do the calculation.

4   There are 1000 grams in a kilogram. Calculate the mass of each shopping basket. Work in kilograms.

  a   3.2 kg of apples, 454 g of jam, 750 g of lentils, 1.2 kg of flour

  b   1.3 kg of sugar, 320 g of strawberries, 0.65 kg of rice

5 Work these out without using a calculator.

    a 4.25×5 **b** 2.84×8 **c** 2.52×6 **d** 1.15×9

    e $18.24 \div 8$ **f** $33.6 \div 7$ **g** $28.4 \div 8$ **h** $17.9 \div 5$

(PS) 6 In an experiment a beaker of water has a mass of 1.104 kg. The beaker alone weighs 0.125 kg. Calculate the mass of water in the beaker, giving your answer in kilograms.

(PS) 7 A piece of string is 5 m long. Pieces of length 84 cm, 1.23 m and 49 cm are cut from it. How much string is left, in metres?

8 A rectangle is 2.35 m by 43 cm.

    a   What is its perimeter, in metres?   **b**   What is its area, in square metres?

(PS) 9 In a 400 m race, William came first with a time of 45.36 seconds, Isaac came second with a speed of 46.14 seconds.

    a   By how many seconds did William win?

    b   How many seconds did it take for each person to run 100 m?

(PS) 10 **a**   Calculate the perimeter of a rectangle that measures 17.3 cm by 11.6 cm.

    b   Calculate the area of the rectangle.

11 Eight cans of orange cost £6.80.

    What is the price of one can?

12 One MP3 player can hold 1.44 GB of information.

    How much information can five MP3 players hold?

13 Ten pencils cost £8.60.

    How much would six pencils cost?

(PS) 14 A can of cola and a toffee bar cost £1.50 together. Two cans of cola and a toffee bar cost £2.35 together.

    How much would four cans of cola and three toffee bars cost?

(MR) 15 A shop sells a range of DVDs at £8.49 each, or £18.90 for three.

    How much would you save if you buy three DVDs together, rather than buying them separately?

(PS) 16 I went into a shop with £10. I wanted to buy three packets of flour at £1.15 each and two bottles of sauce at £2.85 each.

    Will I have enough to buy a box of chocolates costing £2.75 as well?

(PS) 17 Six mugs cost £12.90.

    How much would five mugs cost?

Challenge: Centimetres and millimetres

Centimetres and millimetres both show lengths. The first length shown on the rule below, AB, can be given as 1.6 cm, 16 mm or $1\frac{3}{5}$ cm.

Write each length shown below:

a   in centimetres as a decimal   **b**   in millimetres   **c**   in centimetres as a fraction.

    A   AC   **B**   BD   **C**   CE   **D**   DE   **E**   EF   **F**   EG

Ready to progress?

I can round numbers to the nearest 10 or 100.
I can recall multiplication facts up to 12×12.

I can round numbers to one decimal place.
I can solve problems involving decimals.

I can round numbers to one significant figure.
I can multiply and divide by powers of 10.
I can check answers to problems by estimating the answer.

Review questions

1   Write down the answer to each problem, using words.

    a   Six thousand, four hundred, added to seventeen thousand.

    b   Thirteen thousand, two hundred and twenty, subtract five thousand.

    c   One million, subtract three hundred and thirty-one thousand.

2   Work these out.

    a $9.7 + 5.8$ **b** $9.2 - 3.4$ **c** $3.7 + 4.9 + 2.3$

    d $8.5 - 2.3 - 1.5$ **e** $9.3 - 4.5 - 2.8$ **f** $21 + 2.1 - 1.2$

3   The table shows the approximate populations of five towns.

    a   Which of these places has a population of about eight hundred thousand?

    b   Use the table to complete these sentences.

    The population of . . . . is about twice as big as the population of . . . .

    The population of . . . . is about 4 times as big as the population of . . . .

    The population of Birmingham is about . . . . . . times as big as the population of Hathersage.

Place	Approximate population
Birmingham	1 000 000
Sheffield	800 000
Bradford	500 000
Eastbourne	200 000
Hathersage	1000

ain how you can tell each of the following calculations is wrong.

7 × 15 = 552   **b**   35 × 44 = 1140   **c**   800 ÷ 25 = 33

pay £26.40 to travel to work each week.

work for 46 weeks each year.

How much do I pay to travel to work each year?

how your working.

could buy one season ticket that would let me travel for 46 weeks.

t would cost £966.

How much is that per week?

club is planning a trip. The club hires 234 coaches. Each coach holds 52
assengers. How many passengers is that altogether?

he club wants to put one first aid kit into each of the 234 coaches. The first aid kits
re sold in boxes of 18. How many boxes does the club need?

Anne's height is 1.1 m. Linda is 0.3 m taller than Anne. What is Linda's height?

Kathryn's height is 1.48 m. Eve is 0.7 m shorter than Kathryn. What is Eve's height?

CDs are 45p each. How much will a box of 35 CDs cost. Give your answer in
ds.

an buys 18 bottles of lemonade for a party.

uys the bottles in packs of 6.

ts £4.80 for each pack.

non also buys 18 bottles of lemonade for the same party.

uys the bottles in packs of 3.

ts £2.70 for each pack.

Who paid more for the 18 bottles?

How much more did this person pay?

d each number to one significant figure.

38   **b**   4871   **c**   71.9   **d**   0.8026
0742   **f**   156   **g**   57.8   **h**   2540

Challenge

Space – to see where no one has seen before

You can see objects in space which are a staggering 14 billion light years away. A light year is the distance that light travels in one Earth year. Because it takes so long for light to reach us from distant objects we do not even know whether they still exist by the time we see them!

Although we have not yet seen anything further away than 14 billion light years we think that the diameter of the universe where we can observe objects if they exist is 93 billion light years. We think of this observable universe as a sphere with the observers (us) at the centre. So its edge is 46.5 billion light-years away from us.

1 Light travels at approximately 300 000 000 metres per second.
   Work out the distance light will travel in a year (a light year).

2 Find the furthest distance, in kilometres that we can observe in space.

3 Rewrite the information giving each figure to one significant figure.
   Professor Ball estimated that there are 152 317 298 405 stars in our galaxy
   He also estimated that there are 163 724 308 912 galaxies in the universe

4 Give the number of stars in our galaxy to 2 significant figures.

5 Give the number of galaxies in the universe to 3 significant figures.

6 Assume all galaxies have a similar number of stars. How many stars are there
   altogether in the universe? Give your answer to 2 significant figures.

The radius of our Sun, which is the closest star to us, is 700 000 km.
The radius of the Earth is 6400 km.

The formula for finding the approximate volume of a sphere is $V = 4 \times (\text{radius})^3$

To find the volume of a sphere with a radius of 8 cm:

volume $= 4 \times 8^3$
$\qquad = 4 \times 8 \times 8 \times 8$
$\qquad = 2048 \text{ cm}^3$
$\qquad = 2050 \text{ cm}^3$ (3 sf)

7 Calculate the volume of the Earth.

8 Calculate the volume of the Sun.

9 How many Earths could you fit into the sun?

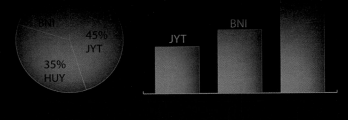

Distribution of the securities market key players

erpreting
a

Distribution of the securities market key players

chapter is going to show you:

- to interpret pie charts
- to create simple pie charts
- to read information from different charts and rams.

should already know:

- to interpret simple data from tables, graphs and ts
- to find the mode, median and range for small sets.

t this chapter

standing data can help you to see what to do, ow and in the future, and sometimes what not You will need to make sense of the huge amount a presented to you on TV and via the internet, apers and magazines. Then you should look at the sions other people draw from data and decide er you agree with them.

ay that data is represented can make it misleading. know how to spot this, it will help you to stand how to present data yourself, to show sions or to make a point clearly.

chapter you will look at some commonly used of statistical diagrams – pie charts, line graphs catter graphs – and learn how to interpret them tly and create them yourself.

PS

9.1 Information from charts

Learning objective

- To revise reading from charts and tables

You often see charts in magazines and newspapers, giving a lot of different types of information.

Example 1

Here are the results of a girls' long-jump competition.

The table shows how far they jumped, in centimetres.

Person	1st jump	2nd jump	3rd jump	4th jump
Amy	218	105	233	297
Donna	154	108	287	176
Gaynor	202	276	95	152
Joy	165	197	240	295

a How far did Gaynor jump on her third jump?

b Who improved with every jump?

c Who won the competition?

   All the information is found by reading the table.

   a Gaynor jumped 95 cm on her third jump.

   b Joy was the only girl who improved with every jump.

   c Amy had the longest jump of 297 cm, so she won the competition.

Exercise 9A

 1 The chart shows the distances, in miles, between certain towns in the UK.

Birmingham				
159	Exeter			
144	302	Holyhead		
206	365	247	Newcastle	
76	235	149	129	Sheffield

a How far apart are Exeter and Newcastle?

b How far apart are Birmingham and Holyhead?

c Which town is 235 miles from Exeter?

d Which two towns are 206 miles from each other?

e How much further away from Newcastle is Exeter than Sheffield?

2 The pictogram shows how many DVDs five pupils have in their collections.

Callum

Jack

Emily

Mia

Raj

Key

represents 4 DVDs

a Who has the most DVDs?

b How many DVDs does Mia have?

c How many DVDs does Jack have?

d How many more DVDs does Emily have than Raj?

e How many DVDs do the five pupils have altogether?

3 The dual bar chart shows the daily mean number of hours of sunshine in London and Bristol over a year.

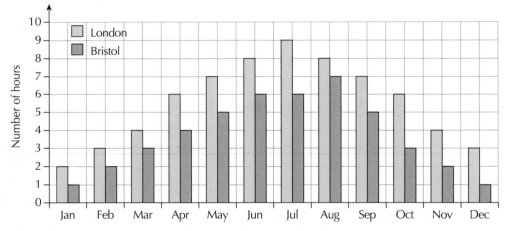

a Which city has the most sunshine?

b Which month is the sunniest for:

i London ii Bristol?

c What is the range for the number of hours of sunshine over the year for:

i London ii Bristol?

4 The line graph shows the temperature, in degrees Celsius (°C), in Birmingham over a 12-hour period.

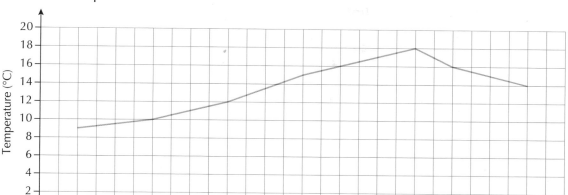

a What was the temperature at midday?

b What was the temperature at 3:00 pm?

c Write down the range for the temperature over the 12-hour period.

d Explain why the line graph is a useful way of showing the data.

5 This compound bar chart shows the favourite colours for a sample of Year 8 pupils.

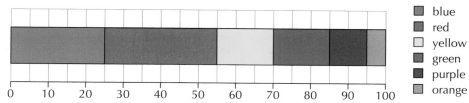

blue
red
yellow
green
purple
orange

a Which colour was chosen by the greatest number of pupils?

b What percentage of the pupils chose yellow?

c Which two colours were equally liked by the pupils?

d Explain why the compound bar chart is a useful way to illustrate the data.

6 The bar chart shows the marks obtained in a mathematics test by the pupils in class 8KG.

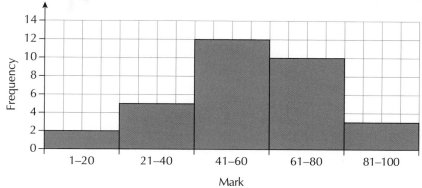

a How many pupils are there in class 8KG?

b How many pupils got a mark over 60?

c Write down the smallest and greatest range of marks possible for the data.

9.2 Reading pie charts

Learning objective

- To interpret a pie chart

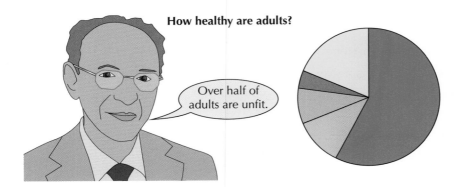

How healthy are adults?

Over half of adults are unfit.

In the pie chart, which colour represents 'Unfit adults'? How do you know?

The pie chart is used because it shows the proportion of the whole amount and is quite easy to interpret.

Sometimes you will have to read information from pie charts, and sometimes you will be asked to draw them.

In a pie chart, the information is represented by a whole circle (a pie) and each category is represented by a **sector** of the circle (a slice of the pie).

Example 2

This pie chart shows the proportion of British and foreign cars sold one weekend at a car showroom. 40 cars were sold.

a How many British cars were sold?

b How many foreign cars were sold?

You can see from the pie chart that $\frac{1}{4}$ of the cars sold were British and $\frac{3}{4}$ were foreign. So:

a $\frac{1}{4}$ of 40 = 10: 10 British cars were sold.

b $\frac{3}{4}$ of 40 = 30: 30 foreign cars were sold.

Cars sold one weekend

British cars

Foreign cars

Example 3

The pie chart shows how one country dealt with 3000 kg of dangerous waste in 2013.

Disposal of dangerous waste, 2013

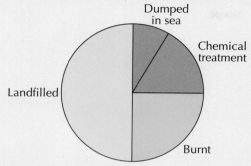

How much waste did they destroy by:

a putting it into landfill   **b** burning it

c dumping it at sea   **d** chemical treatment?

From the pie chart you can see that:

a $\frac{1}{2}$ of the waste was put into landfill.

$\frac{1}{2}$ of 3000 kg = 1500 kg

So 1500 kg of the waste was put into landfill.

b $\frac{1}{4}$ of the waste was burnt.

$\frac{1}{4}$ of 3000 = 3000 ÷ 4 = 750

So 750 kg of the waste was burnt.

c The angle of this sector is 30°.

Then the fraction of the circle is:

$$\frac{30}{360} = \frac{1}{12}$$

Therefore $\frac{1}{12}$ of the waste was dumped at sea.

3000 ÷ 12 = 250

So 250 kg of the waste was dumped at sea.

d The angle of this sector is 60°.

Then the fraction of the circle is:

$$\frac{60}{360} = \frac{1}{6}$$

Therefore $\frac{1}{6}$ of the waste was treated by chemicals.

3000 ÷ 6 = 500

So 500 kg of the waste was treated by chemicals.

1 600 pupils in a school were asked to vote for the type of film they would be shown at Christmas.

The pie chart illustrates their responses. How many voted for:

a comedy   **b** romance   **c** action?

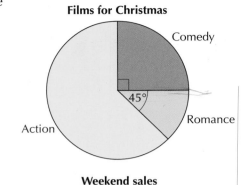

Films for Christmas

2 One weekend in Castleton a fish and chip shop took £3000 in sales.

The pie chart shows the proportions of money taken on each type of food.

How much was taken that weekend on:

a sausages   **b** pies
c fish   **d** chips?

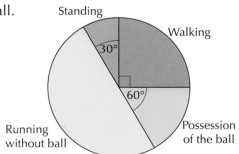

Weekend sales

3 Ali had 240 friends on her social media account.

She asked them all if they eat fruit.

The pie chart illustrates her results.

How many of these friends:

a never eat fruit

b sometimes eat fruit

c eat fruit every day?

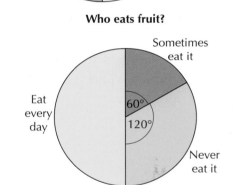

Who eats fruit?

MR **4** The pie chart shows how long Roy Rogers spent on each different aspect during a game of football.

A game lasts for 90 minutes.

a How long did Roy spend:

i in possession of the ball

ii walking

iii running without the ball

iv standing?

b The manager said: 'Roy was in possession of the ball for over one sixth of the game.'

Was he correct? Explain your answer.

Activities in a football match

5 A plumber was explaining how much time he spent on different aspects of installing a new bathroom.

The pie chart illustrates how he broke his time down for a job that was to last nine days, working for 8 hours each day.

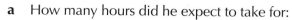

a How many hours did he expect to take for:

   i gutting the bathroom

   ii relaying all pipework

   iii electrical work

   iv installation of fittings?

b The plumber said: 'Half my time will be spent tiling.'

Was he correct? Explain your answer.

6 One week there were 180 days absence at a school. The pie chart illustrates how many pupils were absent on each day.

a How many pupils were absent on each day?

   i Monday   **ii** Tuesday   **iii** Wednesday   **iv** Friday

b The headteacher reported that one third of the absences were on the Friday.

Is this correct? Explain your answer.

c Explain how you can tell what simple fraction was absent on the Thursday.

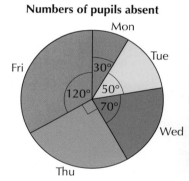

Activity: Posters

Design a poster to show information about 20 pupils in your class. Either include pie charts that you have drawn yourself or use a spreadsheet to produce the pie charts. Make sure that any pie chart you produce has labels and is easy to understand.

9.3 Creating pie charts

Learning objective

* To use a scaling method to draw pie charts

Key words	
frequency	frequency table
scaling	total frequency

Sometimes you will have to draw a pie chart to display data that is given in a **frequency table**.

Example 4

Draw a pie chart to represent this set of data, showing how a group of people travel to work.

Type of travel	Walk	Car	Bus	Cycle
Frequency	30	120	60	30

Each type of travel (walk, car, bus and so on) will be represented by its own sector in the pie chart.

The size of the sector will depend on the **frequency** for that sector.

The **total frequency** is 240 people.

A pie chart has an angle of 360° at its centre.

To find the angle that represents each frequency, you need to work out the frequency as a fraction of the total, then multiply that by 360°.

This is called the **scaling** method.

The angle for each sector is:

$$\frac{\text{frequency for that sector}}{\text{total frequency}} \times 360°$$

Set the data out in a vertical table and write the calculations in it.

Sector (type of travel)	Frequency	Calculation	Angle
Walk	30	$\frac{30}{240} \times 360° = 45°$	45°
Car	120	$\frac{120}{240} \times 360° = 180°$	180°
Bus	60	$\frac{60}{240} \times 360° = 90°$	90°
Cycle	30	$\frac{30}{240} \times 360° = 45°$	45°
Total	240		360°

Now draw the pie chart.

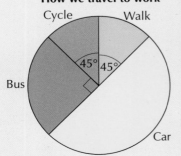

Hint When drawing a pie chart, draw the smallest angle first and try to make the largest angle the last one you draw, then any cumulative error in drawing will not be so noticeable.

Exercise 9C

1 Draw a pie chart to represent the types of food that 40 people usually eat for breakfast.

Food	Cereal	Toast	Other	None
Frequency	20	5	10	5

2 Draw a pie chart to represent the favourite subjects of 36 pupils.

Subject	Maths	English	Science	Other
Frequency	12	9	9	6

3 Draw a pie chart to represent the numbers of goals scored by an ice-hockey team 24 matches.

Goals	0	1	2	3	4	5 or mo
Frequency	2	6	6	4	4	2

4 Draw a pie chart to represent the favourite colours of 60 Year 8 pupils.

Colour	Red	Green	Blue	Yellow	Other
Frequency	15	10	20	5	10

(MR) 5 This is an estate agent's waiting list for rental flats on one day in a large city.

Type of flat	Number of applicants
1 bedroom	70
2 bedrooms	200
3 bedrooms	80
4 bedrooms	7
5 bedrooms	3

a Why would it be difficult to draw a pie chart to illustrate this information?

b Combine the groups so that you could draw a pie chart to represent this informa

c Draw the pie chart.

Challenge: World energy consumption

The bar chart illustrates the world energy consumption, given as percentages of the total energy used.

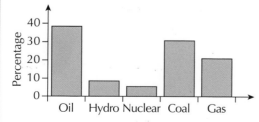

A Find the percentage, as accurately as you can, for each type of energy.

B Explain why your total may not come to 100%.

C Draw a pie chart to represent this data.

4 Scatter graphs

rning objective

ead scatter graphs

tor recorded the brightness of the sunlight and the size of the pupils of people's eyes.

en plotted the results on a graph. The result is a scattering of points on the graph, which is why pe of graph is called a **scatter graph**.

can you tell about the connection between the brightness of the sunlight and the size of people's
?

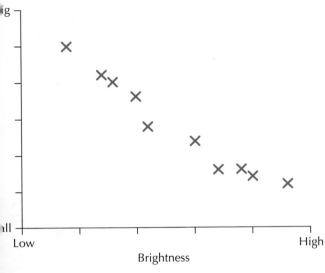

an see that as the light gets brighter the size of people's pupils gets smaller.

ample 5

scribe the relationships in each scatter graph.

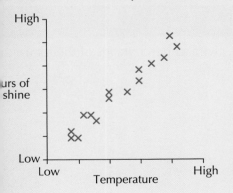

This graph shows that the temperature is higher when there are more hours of sunshine.

9 Interpreting data

b

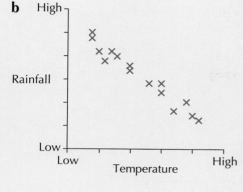

This graph shows that as the temperature increases, the rainfall decreases.

c

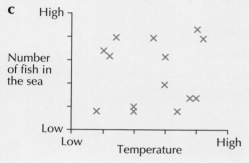

This graph shows that there is no connection between the temperature and the number of fish in the sea.

Exercise 9D

  1 After completing a study into temperature, rainfall and the sales of ice creams, umbrellas and the hire of deckchairs, a tour guide drew these scatter graphs. Copy and complete the sentences to describe the link.

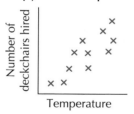 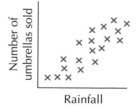

a The number of deckchairs increases as the temperature ....

b The number of deckchairs hired decreases as the rainfall ....

c The number of umbrellas sold .... as the rainfall increases.

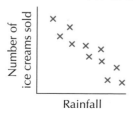 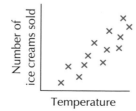 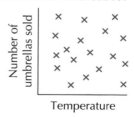

d The number of ice creams sold increases as the rainfall ....

e The number of ice creams sold increases as the .... increases.

f There is .... between temperature and number of umbrellas sold.

2 After completing a study into height, mass, shoe size and collar size, a medical student drew these scatter graphs. Copy and complete the sentences about the connection each graph shows.

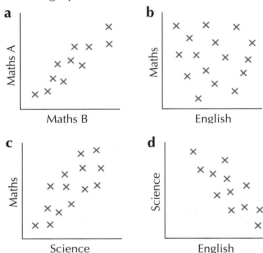

a The height of a person increases as the shoe size ....

b There is no relationship between mass and ....

c The mass of a person increases as the collar size ....

d There is .... between height and collar size.

3 After a series of mathematics, science and English test results, a teacher drew these scatter graphs. Describe the connection shown on each graph.

a

Maths A vs Maths B

b

Maths vs English

c

Maths vs Science

d

Science vs English

 4 A fan surveyed the transfer price and age of some goalkeepers, as well as how many goals they let in during their first full season. She drew these scatter graphs. Describe what each graph tells you.

a **Premiership**

Number of goals let in vs Price

b **Premiership**

Number of goals let in vs Age

c **League 2**

Number of goals let in vs Price

d **League 2**

Number of goals let in vs Age

5 These three scatter graphs show the heights of people and the cost of their clothes.

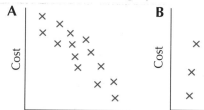

Billy said: 'The shorter you are, the less your clothes will cost.'

Terry said: 'The shorter you are, the more your clothes will cost.'

Suzie said: 'Your height doesn't make any difference to the cost of your clothes.'

Match up the correct graph with the statements each person has made.

6 Mr Salomon set his class two mathematics tests.

When he marked them, he created a scatter graph to compare the results.

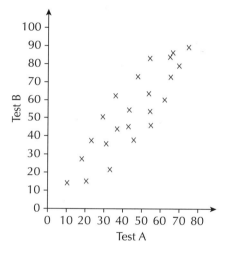

a What can you say about the results of the two tests?

b Joe was absent for the tests, but when he came back he sat test A and scored 75 marks. Approximately how many marks might you expect him to get on test B?

c Lilly was also absent from the tests. but took Test B when she came back. She scored 10 marks. How many marks might you expect her to get if she took test A?

Activity: Correlation in circles

Try to find out if there is any connection between the diameter of a circle and its circumference.

- Use a variety of circular objects, such as tins and coins.
- Use compasses to draw some circles.
- Measure the diameter and the circumference of each circle and record your measurements.

Is there a pattern?

Ready to progress?

I can read information from bar charts and pie charts.

I can create simple pie charts.
I can read information from scatter graphs.

Review questions

1   The table shows the average length of pregnancy for different animals.

Mammal	Average length of pregnancy
sheep	150 days
cow	284 days
donkey	374 days
giraffe	365 days
camel	425 days
black rhino	465 days

Use the information in the table to answer these questions.

a   Which animal has an average length of pregnancy of about 1 year?

b   Which animal has an average length of pregnancy of about 21 weeks?

c   A human has an average length of pregnancy of about 9 months. Which other animal also has an average length of pregnancy of about 9 months?

2   a   Helen asked 25 people: 'Do you like pizza?'

The bar chart shows her results for 'Yes' and 'No'.

Copy the bar chart and complete it to show her results for 'Don't know'.

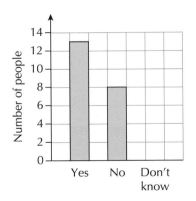

b   This pictogram also shows her results for 'Yes' and 'No'.

Copy the pictogram and complete it to show her results for 'Don't know'.

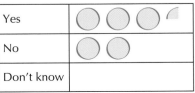

Key represents 4 people

3 Abby wanted to find out why people in some countries lived longer than people in other countries.

She picked 15 countries and found out:

- the population
- the number of doctors per million people
- the average length of life.

She plotted these graphs to help her look for links:

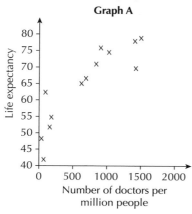

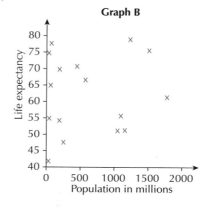

a What does graph A show about possible links between length of life and the number of doctors per million people in a country?

b What does graph B show about possible links between length of life and a country's population?

c Abby is told that another country has:

- a population of about 100 million
- about 800 doctors per million people.

Use the graphs above to estimate the average length of life for this country.

4 There are 24 pupils in Mr Lockwood's class.

He did a survey of how the pupils in his class travelled to school.

He started to draw a pie chart to show his results.

a Four of his pupils travel by car.

   i Sketch a copy of the pie chart, showing these four pupils.

   ii The only pupils now not shown are those who travel by taxi. How many pupils travel by taxi?

b There are 36 pupils in Mrs Casey's class.

She did the same survey and drew a pie chart to show her results.

How many of Mrs Casey's pupils travelled by:

   i taxi   **ii** bicycle   **iii** car?

c Mr Lockwood said: 'Mrs Casey's chart shows fewer pupils travelling by bus than mine does.'

He is wrong. Explain why.

How pupils travel to school

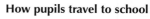

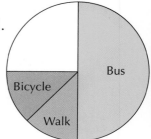

How pupils in Mrs Casey's class travel to school

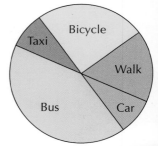

Challenge
What should we eat?

We are given guidance by experts on what food is good for us and how much of it we should eat.

For many years the guidance was to eat 5 portions of fruit or vegetables a day.

Recently, the same experts said we should be eating 7 portions a day.

Some even believe we should be eating 10 portions a day!

Here are some facts from a report completed in 2014.

- You can reduce the risk of an early death by the number of portions of fruit and vegetables you eat.
- 1 to 3 portions a day reduces risk by 15%.
- 4 or 5 portions a day reduces risk by 30%.
- 6 or 7 portions a day reduces risk by 35%.
- Over 7 portions a day reduces risk by 40%.

1 Draw a bar chart to illustrate the figures in the box on the opposite page.

In the UK we eat an average of 350 g of fruit and vegetables a day.
It is recommended that we should eat 400 g of fruit and vegetables a day.

2 Draw a pie chart to illustrate how much of our recommended portion we actually eat in the UK.

Current data suggests that the average annual consumption of fruit and vegetables in the UK is about 125 kg per person. In the USA, the average annual consumption is about 135 kg per person.

3 Draw a chart to illustrate these facts.

It is said that 70% of people in the UK are not eating enough fruit and vegetables.

4 a Assume the population of the UK to be 64 million people.
   How many are eating enough fruit and vegetables?

   b Write a report to explain what needs to happen for 50% of the UK to be eating enough fruit and vegetables.

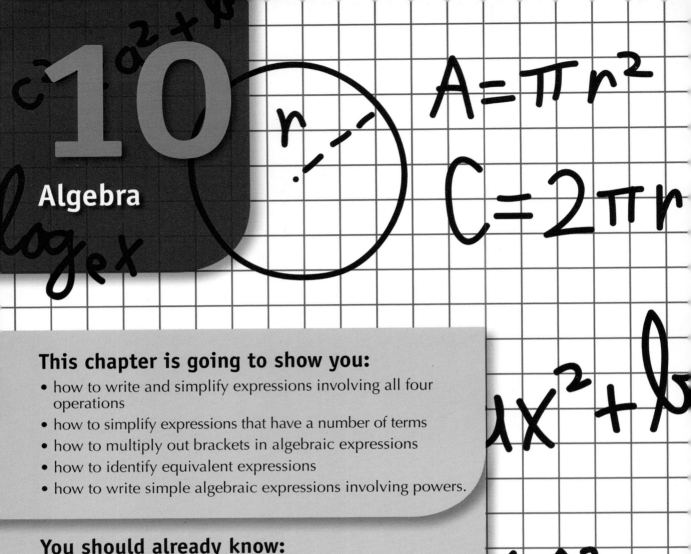

10

Algebra

$$A = \pi r^2$$

$$C = 2\pi r$$

This chapter is going to show you:
- how to write and simplify expressions involving all four operations
- how to simplify expressions that have a number of terms
- how to multiply out brackets in algebraic expressions
- how to identify equivalent expressions
- how to write simple algebraic expressions involving powers.

You should already know:
- what an algebraic expression is
- how to interpret simple algebraic expressions
- how to write simple algebraic expressions.

About this chapter
When the bar code of an item is scanned at the till in a shop, a series of calculations is carried out on all the digits except the last one. The calculations give a remainder, which is the final digit in the bar code. If the two numbers do not match, the bar code has not been scanned correctly.

The series of calculations can be described using algebra.

Algebra is very useful in science, and in areas such as computer programming, because it helps us to write things simply and clearly. The rules of algebra are like the grammar of a language. If you know them, you can read algebraic sentences and write them yourself. This chapter is about some of the rules of algebra.

10.1 Algebraic notation

Learning objective

* To simplify algebraic expressions involving the four basic operations

Key word

algebraic expression

You already know that you usually leave out the multiplication sign (×) when you write an **algebra** **expression**.

For example, instead of $4 \times a$ or $a \times 4$, you should write $4a$.

Notice that you always put a number in front of a letter. Never write it as $a4$. You will see why this important, in section 10.5.

You do not usually use the division sign (÷) either.

$a \div 4$ is written as $\frac{a}{4}$.

You can also use a fraction to write $a \div 4$ as $\frac{1}{4}a$. This is because to find a quarter of a number you divide it by 4. The two operations give the same answer.

Example 1

Write each of these as simply as possible.

a $2 \times t$ **b** $k \times 3 + 5$ **c** $d \times (7 - 2)$

a $2 \times t = 2t$	You can leave out the × sign.
b $k \times 3 + 5 = 3k + 5$	Put the 3 in front of the k.
c $d \times (7 - 2) = d \times 5$	Work out the subtraction in the brackets first.
$d \times 5 = 5d$	Put the 5 in front of the d.

If you have more than two things to multiply, you can do them in any order.

When a letter is multiplied by itself, use the 'square' notation to write it down.

For example, write $d \times d$ as d^2.

Example 2

Write each of these as simply as possible.

a $p \times k$ **b** $2 \times k \times 4$ **c** $f \times 5 \times f$

a $p \times k = pk$	You do not have to put the letters in alphabetical order.
b $2 \times k \times 4 = 8k$	Multiply the numbers first.
c $f \times 5 \times f = 5f^2$	Put the 5 in front and then $f \times f = f^2$.

1 Write these expressions as simply as possible.

a $3 \times p$ b $t \times 5$ c $4 \times x$ d $y \times 6$

e $a \times b$ f $n \times m$ g $p \times 5$ h $m \times 7$

i $6 \times t$ j $5 \times e$ k $3 \times q$ l $9 \times t$

m $a \times t$ n $n \times q$ o $p \times 7$ p $p \times 6$

2 Write down the pairs of expressions that are equal to each other.

The first one has been done for you.

a $a + b$ $b + a$ ab

$\boxed{a + b = b + a}$

b $m + p$ $p + m$ mp

c $a - b$ $b - a$ $-b + a$

d $m \div n$ $n \div m$ $\dfrac{n}{m}$

e $4 \times b$ $4 + b$ $4b$

3 Write each expression as simply as possible.

a $a \times 4p$ b $3t \times 5$ c $4m \times x$ d $2y \times 3$

e $3a \times b$ f $4n \times m$ g $m \times 5q$ h $3p \times 7$

i $5q \times p$ j $4 \times 3d$ k $2q \times p$ l $8x \times y$

m $2a \times 5b$ n $5n \times 5m$ o $3p \times 7q$ p $3p \times 6m$

4 Write each expression as simply as possible.

a $4 \times a \times 2$ b $3 \times p \times 4$ c $3 \times 6 \times t$ d $3 \times 10 \times k$

e $a \times 4 \times 5$ f $M \times 5 \times 2$ g $4 \times x \times 5$ h $3 \times 4 \times t$

5 Write these expressions as simply as possible.

a $2 \times m \times n$ b $5 \times p \times q$ c $3 \times t \times p$ d $5 \times b \times t$

e $m \times 3 \times q$ f $n \times 6 \times m$ g $2 \times x \times y$ h $b \times 4 \times t$

6 Write these expressions as simply as possible. The first one has been done for you.

a $a \times a = a^2$ b $x \times x$ c $3 \times m \times m$ d $q \times q \times 4$

e $2 \times t \times t$ f $e \times 5 \times e$ g $4 \times t \times t$ h $3 \times n \times n$

7 Write each expression as simply as possible.

a $3 \times 2m$ b $5 \times 4t$ c $2 \times 5p$ d $6 \times 2n$

e $4 \times 3x$ f $5k \times 2$ g $4p \times 5$ h $2 \times 6t$

8 Write each expression without using a division ($\div$) sign. The first one has been done for you.

a $x \div 3 = \dfrac{x}{3}$ b $m \div 5$ c $n \div 4$ d $k \div 2$

9 Only some of these statements are true.

State which are true and which are false.

a $a + b = c + d$ is the same as $c + d = a + b$

b $m - n = 7$ is the same as $7 = m - n$

c $6t = m + 4$ is the same as $m = 6t + 4$

d $4 - 3n = 9$ is the same as $9 = 3n - 4$

e $bt - pq = A$ is the same as $A = bt - pq$

  10 **a** Explain why $a - b$ does not always equal $b - a$.

b Give two examples of where $a - b$ will equal $b - a$.

Challenge: Is it true?

Some of these statements are always true, whatever values you choose for m and p.

a $m + p + 1 = 1 + m + p$

b $m + 8 - p = m - 8 + p$

c $m \times p + 3 = 3 + m \times p$

d $\dfrac{6 + m}{2} = 3 + \dfrac{m}{2}$

A Which ones are always true?

B Explain how you know that the others are not true.

10.2 Like terms

Learning objective

- To simplify algebraic expressions by combining like terms

Key words

| like terms | term |

Look at this expression:

$5a + 4b - 3a + b$

It has four **terms** altogether. There are two terms that include a and two terms with b.

You can write the terms in a different order like this:

$5a - 3a + 4b + b$

You can combine the a-terms and you can combine the b-terms.

$5a - 3a = 2a$ and $4b + b = 5b$

So $5a - 3a + 4b + b = 2a + 5b$.

$5a$ and $3a$ are called **like terms** and like terms can be combined to simplify an expression.

$4b$ and b are also like terms.

$2a$ and $5b$ are not like terms, because they contain different letters.

The expression $2a + 5b$ cannot be simplified any further.

Example 3

Simplify each expression as much as possible.

a $4f - 3f$ **b** $4m + 2m - 3$ **c** $2g - 3h + 2g - 4h$ **d** $x^2 + 4x + x - 3$

   a $4f - 3f = f$ The two terms are like terms and can be combined.

       Notice that you write f not $1f$.

   b $4m + 2m - 3 = 6m - 3$ The first two terms are like terms and can be combined.

       The 3 and the m are not like terms and the expression cannot be simplified further.

   c $2g - 3h + 2g - 4h = 2g + 2g - 3h - 4h$ Put like terms together.

                  $= 4g - 7h$ Combine the g-terms and the h-terms.

       Notice that subtracting $3h$ and then subtracting another $4h$ is the same as subtracting $7h$.

   d $x^2 + 4x + x - 3 = x^2 + 5x - 3$ Only the middle terms are similar and can be combined.

Exercise 10B

1   Simplify each expression.

    a $m + m$ **b** $k + k + k$ **c** $a + a + a + a$ **d** $d + d + d$

    e $q + q + q + q$ **f** $t + t$ **g** $n + n + n + n$ **h** $g + g + g$

    i $p + p + p$ **j** $w + w + w + w$ **k** $i + i + i + i + i$ **l** $a + a + a + a$

2   Copy and complete each statement. The first one has been done for you.

    a $t + t + t + t = 4 \times t = 4t$ **b** $p + p + p = 3 \times p = \ldots$

    c $m + m + m + m = 4 \times m = \ldots$ **d** $k + k + k = \ldots\ldots = \ldots$

    e $h + h + h + h + h = \ldots\ldots = \ldots$ **f** $\ldots\ldots\ldots = 6 \times m = 6m$

    g $\ldots\ldots\ldots = 5 \times p = \ldots$ **h** $\ldots\ldots = 3 \times g = \ldots$

    i $\ldots\ldots\ldots = \ldots\ldots = 7n$ **j** $\ldots\ldots\ldots = \ldots\ldots = 5y$

    k $\ldots\ldots\ldots\ldots = \ldots\ldots = 8p$ **l** $\ldots\ldots\ldots = \ldots\ldots = 10t$

3   Simplify each of these.

    a $5g + g - 2g$ **b** $3x + 5x - 6x$ **c** $4h + 3h - 5h$ **d** $4q + 7q - 3q$

    e $5h - 2h + 4h$ **f** $6x - 4x + 3x$ **g** $3y - y + 4y$ **h** $5d - 4d + 6d$

    i $8x - 2x - 3x$ **j** $5m - m - 2m$ **k** $8k - 3k - 2k$ **l** $6n - 3n - n$

4   From each cloud, group together the like terms. For example:

    a $3b$ $2a$ $\longrightarrow$ $a, 2a, 5a$

    $4b$ $5a$ b $b, 3b, 4b$

a

    $3t$ g $5t$

    $8g$ $9t$ $7g$

b

    m $7p$ $4m$

    $9p$ $10m$ $3p$

c

    $4k$ $3m$ $5w$

    $8m$ $7w$ $7m$ k

d

    x^2 t $5x^2$

    $3t$ $3x^2$ $4t$

e

    y^2 $2y$ $8y$

    $7y^2$ $4y^2$ $3y$

f

    $7w$ $7g$ $3h$ $3w$

    $3g$ $9h$ $4w$ $10g$

5 Simplify each of these expressions.

a $3b + 5 + 2b$ **b** $2x + 7 + 3x$ **c** $m + 2 + 5m$ **d** $4k + 3k + 8$

e $3x + 7 - x$ **f** $5k + 4 - 2k$ **g** $6p + 3 - 2p$ **h** $5d + 1 - 4d$

i $5m - 3 - 2m$ **j** $6t - 4 - 2t$ **k** $4w - 8 - 3w$ **l** $5g - 1 - g$

m $t + k + 4t$ **n** $3x + 2y + 4x$ **o** $2k + 3g + 5k$ **p** $3h + 2w + w$

q $5t - 2p - 3t$ **r** $6n - 2t - 5n$ **s** $p + 4q - 2q$ **t** $3n + 2p - 3n$

6 Simplify each of these expressions.

a $2t + 3g + 5t + 2g$ **b** $4x + y + 2x + 3y$ **c** $2m + k + 3m + 2k$

d $5x + 3y - 2x + y$ **e** $6m + 2p - 4m + 3p$ **f** $3n + 4t - n + 3t$

g $6k + 3g - 2k - g$ **h** $7d + 4b - 5d - 3b$ **i** $4q + 3p - 3q - p$

j $4g - k + 2g - 3k$ **k** $2x - 3y + 5x - 2y$ **l** $4d - 3e - 3d - 2e$

7 Simplify each expression.

a $8e + 5 + 6e + 3$ **b** $9x - 3 + 7x - 1$ **c** $8 + 3a + 5 - 2a$

d $4 + 4d + 3 + 3d$ **e** $2 + 3h - 2h + 4$ **f** $7k + 5 - 5k - 3$

g $8p + 7 - 5p - 1$ **h** $4k - 5 + 3k + 8$ **i** $5m - 4 + 4m - 2$

j $1 + 6p - 8p + 4$ **k** $7t - 3 - 5t + 5$ **l** $10 + 6q - 3 - 8q$

8 Simplify each expression.

a $2bt + 5bt$ **b** $3mp + mp$ **c** $8pq - 3pq$ **d** $4ab - ab$

e $3kh + 6kh$ **f** $7pt - pt$ **g** $8mr - 5mr$ **h** $8xy - 2xy$

i $2km + m + 4m$ **j** $9mt + 2t - t$ **k** $8dg - 2d + 3d$ **l** $4xy + 5xy - x$

m $8t + 4t - 5pt$ **n** $3x - 4xy + 7x$ **o** $8mn + 3mn + n$ **p** $3h - 5hk + 7hk$

9 The expression in each box is made by adding the expressions in the two boxes it stands on. Copy the diagrams and fill in the missing expressions.

a

?
$3x + 4y$	$5x + 3y$

b

?
$2p + 6t$	$3p + 2t$

c

$3n + 5c$
$2n + c$	?

d

$6a + 4b$
?	$4a + b$

Investigation: Consecutives

A Explain why the sum of two consecutive integers added together is always odd.

B Explain why two consecutive integers multiplied together always give an even number.

C Show that the sum of any three consecutive integers will always give a multiple of 3.

10.3 Expanding brackets

Learning objective

- To remove brackets from an expression

Key words

| expand | multiply out |

You have seen how brackets can be used in algebraic expressions. In this section you will learn how to remove brackets and write expressions in different ways.

Look at these multiplications.

$3 \times (5 + 2) = 3 \times 7 = 21$

$3 \times 5 + 3 \times 2 = 15 + 6 = 21$

This shows that:

$3 \times (5 + 2) = 3 \times 5 + 3 \times 2$

You get the same result if you multiply 5 and 2 by 3 separately and add the results.

A diagram may make this clearer.

$\times$	5	+ 2
3	15	+6

$15 + 6 = 21$

This is a general result. This means that it is always true, even if you replace one or more of the numbers by letters.

Here are some more examples.

- $2(a + 3) = 2a + 6$ This means $2 \times a$ added to 2×3.

$\times$	a	+ 3
2	$2a$	+6

- $c(d - 4) = cd - 4c$ This means $c \times d$ take away $c \times 4$.

$\times$	d	− 4
c	cd	−4c

This is called **multiplying out** or **expanding** an expression with brackets.

Example 4

Write each expression without brackets, as simply as possible.

a $3(t - 5)$ **b** $m(p + 4)$ **c** $2(a + 3) + 3(a - 4)$

  a $3(t - 5) = 3t - 15$ That means $3 \times t$ minus 3×5.

  b $m(p + 4) = mp + 4m$ $m \times p = mp$ and $m \times 4 = 4m$.

  c $2(a + 3) = 2a + 6$ Multiply out the brackets separately.

    $3(a - 4) = 3a - 12$

    So $2(a + 3) + 3(a - 4) = 2a + 6 + 3a - 12$ Put the two expressions together.

                $= 2a + 3a + 6 - 12$ Put like terms together.

                $= 5a - 6$ $2a + 3a = 5a$ and $6 - 12 = -6$

Exercise 10C

1. Calculate each of the following by:
   - **i** expanding the bracket
   - **ii** calculating the bracket first.

   The first one has been done for you.

   a $2(5 + 3)$
   - **i** $2(5 + 3) = 2 \times 5 + 2 \times 3 = 10 + 6 = 16$
   - **ii** $2(5 + 3) = 2 \times 8 = 16$

   b $3(4 + 2)$ **c** $5(3 + 1)$ **d** $4(2 + 3)$ **e** $6(3 + 4)$ **f** $8(5 - 2)$

2. Multiply out each expression.

   a $2(y + 4)$ **b** $4(2p + 3)$ **c** $3(4t + 1)$ **d** $2(3m + 3)$

   e $4(6 + 2n)$ **f** $3(1 + 4h)$ **g** $6(2 + 5k)$ **h** $5(2 + 3p)$

   i $3(4t - 2)$ **j** $2(5 - 2t)$ **k** $8(2 - 3k)$ **l** $3(5 - 2a)$

   m $2(4e + 3d)$ **n** $5(3m + 2)$ **o** $2(4t + 3p)$ **p** $7(3p + 4q)$

   q $2(4k - 3h)$ **r** $3(2x - 4y)$ **s** $4(6m - 4n)$ **t** $5(3p - 2t)$

3. Expand and simplify each expression.

   a $9t + 2(4t + 1)$ **b** $7m + 3(2m + 5)$ **c** $6h + 4(3h + 5)$

   d $5k + 3(4 + k)$ **e** $8p + 4(2 + p)$ **f** $5x + 2(3 + 4x)$

   g $7t + 2(5 + 4t)$ **h** $6q + 2(4 + 3q)$ **i** $8y + 3(1 + 4y)$

4. Expand and simplify each expression.

   a $6(2h + 3) - 5h$ **b** $2(4t + 2p) - 3t$ **c** $5(3k + m) - 4m$

   d $5g + 3(3g - 4)$ **e** $8t + 4(5g - t)$ **f** $5m + 3(5m - 4g)$

   g $6m + 2(4m - k)$ **h** $10p + 2(2p - 3m)$ **i** $8h + 2(3h - 4p)$

5. Expand and simplify each expression.

   a $3(2x + 3y) + 2(4x + 2y)$ **b** $2(3p + 2m) + 3(2p + 5m)$

   c $2(5k + 4g) + 4(2k + 3g)$ **d** $2(3e + 2d) + 3(2d + 5e)$

   e $4(5n + 2p) + 2(3n - 4p)$ **f** $3(5t + 3f) + 2(3t - 2f)$

   g $4(p + 6d) + 3(2p - 3d)$ **h** $2(5x - 3y) + 3(4y - x)$

(MR) 6. Oliver and Eve are talking about an expression.

   Explain how both of them can be correct.

Four times $3a$ plus 1 is equal to $12a$ plus 1.

Oliver

No it's not, its equal to $12a$ plus 4.

Eve

10.4 Using algebra

Learning objective

• To use algebraic expressions in different contexts

In mathematics, and in other subjects, you can use algebra to show connections between variables.

Look at the triangle.

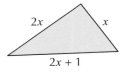

The side lengths are all shown.

We know that:

perimeter = sum of the sides

So in this triangle,

perimeter = $x + 2x + 2x + 1$

$= 5x + 1$

Example 5

Find:

a the perimeter   **b** the area of this rectangle.

    a Perimeter = $2(k + p) = 2k + 2p$

    b Area = $k \times p = kp$

Example 6

State the area of the shape as simply as possible.

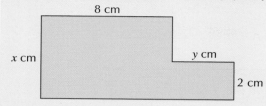

First, split the shape into two parts A and B, as shown.

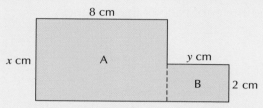

Shape A has the area $8x$ cm^2.

Shape B has the area $2y$ cm^2.

The total area is $(8x + 2y)$ cm^2.

Exercise 10D

1 Write down, as simply as possible, the perimeter of each of these shapes.

a

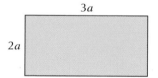

b

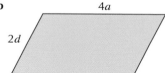

c

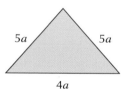

d

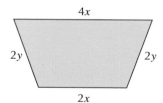

e

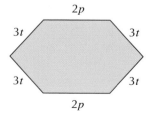

f

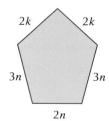

2 What is the area of each rectangle?

a
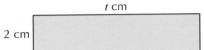
2 cm   *t* cm

b
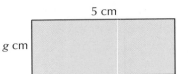
5 cm   *g* cm

c
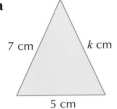
k cm   3 cm

d
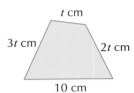
7 cm   *x* cm

3 Write down, as simply as possible, the perimeter of each of these shapes.

a
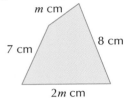
7 cm   *k* cm   5 cm

b

m cm   7 cm   8 cm   2*m* cm

c
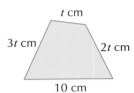
t cm   3*t* cm   2*t* cm   10 cm

4 Work out an expression for the perimeter of each rectangle.
Write your answers as simply as possible.

a
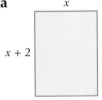
x   *x* + 2

b
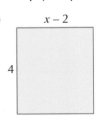
t − 2   *t*

c
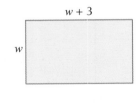
w + 3   *w*

5 Work out an expression for the area of each rectangle.
Write your answers as simply as possible.

a
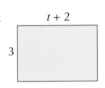
t + 2   3

b

x − 2   4

c
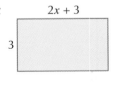
2*x* + 3   3

6 Work out an expression for the perimeter of each shape.
Write your answers as simply as possible.

a
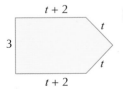
t + 2   *t*   3   *t*   *t* + 2

b
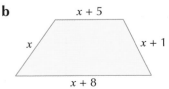
x + 5   *x*   *x* + 1   *x* + 8

c
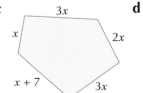
3*x*   *x*   2*x*   *x* + 7   3*x*

d
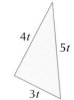
4*t*   5*t*   3*t*

(PS) 7 In this algebra wall, the bricks next to each other are added to give the total in the brick above them.

Show that the expression in the top brick of this algebra wall can be written as $4t + 9$.

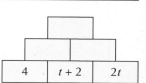

(PS) 8 Show that the expression in the top brick of this algebra wall can be written as $4(t + 2)$.

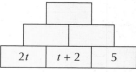

9 The compound shapes below have been split into rectangles.

Work out the area of each rectangle and then add these to get the total area.

Simplify your answers.

a

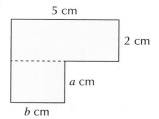

b

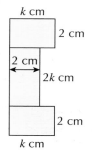

c

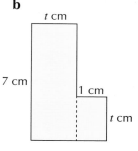

10 In the diagram, all lengths are in centimetres.

a Work out an expression for the area of:

   i rectangle A   **ii** rectangle B

   iii rectangle C   **iv** the whole shape.

b Work out an expression for the perimeter of:

   i rectangle A   **ii** rectangle B

   iii rectangle C   **iv** the whole shape.

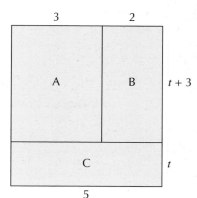

Challenge: Fill in the bricks

Here is an algebra wall.

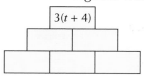

Find three expressions to put in the bottom row of bricks.

10.5 Using powers

Learning objective

- To write algebraic expressions involving powers

You know that you can write $a + a + a$ as $3a$.

In a similar way, you can write $a \times a \times a$ as a^3.

This is called 'a cubed' or 'a to the **power** 3'.

So if the value of a is 4, then:

$a^3 = 4 \times 4 \times 4 = 64$

This is 4 cubed or 4 to the power 3.

You can have powers larger than 3.

$a \times a \times a \times a = a^4$

This is a to the power 4. The number 4 is called the **index**.

Example 7

Write each expression in index form.

a $m \times m \times m \times m$ **b** $n \times n \times n \times n \times n$ **c** $x \times x \times y$ **d** $t(t + 3)$

a $m \times m \times m \times m = m^4$	There are four ms multiplied together.
b $n \times n \times n \times n \times n = n^5$	There are five ns multiplied together.
c $x \times x \times y = x^2 \times y$	You can combine the two xs multiplied together.
$ = x^2 y$	Leave out the multiplication signs.

You could also write it as yx^2.

d $t(t + 3) = t \times t + t \times 3$	Multiply everything in the bracket by t.
$ = t^2 + 3t$	

A very good reason for writing $n \times 3$ as $3n$ and not as $n3$ is so that you will not mistake it for n^3.

Exercise 10E

1 Write each expression in index form.

a $n \times n \times n$ **b** $m \times m$ **c** $p \times p \times p \times p$ **d** $w \times w \times w$

e $m \times m \times m$ **f** $t \times t \times t \times t$ **g** $k \times k \times k \times k$ **h** $y \times y \times y$

i $v \times v \times v \times v$ **j** $d \times d \times d \times d \times d$ **k** $t \times t \times t \times t \times t$ **l** $k \times k \times k$

2 Calculate each of the powers.

a 3^3 **b** 4^2 **c** 2^4 **d** 4^3

e 5^3 **f** 2^3 **g** 3^4 **h** 10^2

i 10^3 **j** 2^5 **k** 8^2 **l** 6^3

3 Write each expression as simply as possible.

a $n \times n$ **b** $m + m$ **c** $p \times p \times p$ **d** $w + w + w$

e $q \times q \times q \times q$ **f** $r + r + r + r$ **g** $k + k$ **h** $f \times f \times f$

i $v + v + v + v$ **j** $d \times d \times d \times d \times d$ **k** $q + q + q$ **l** $t \times t \times t \times t$

4 Copy each of these and write out in full.

a $3t = \ldots$ $t^3 = \ldots$ **b** $4m = \ldots$ $m^4 = \ldots$ **c** $2k = \ldots$ $k^2 = \ldots$

d $5w = \ldots$ $w^5 = \ldots$ **e** $3d = \ldots$ $d^3 = \ldots$ **f** $6p = \ldots$ $p^6 = \ldots$

5 Write each expression in index form.

a $n \times n \times p$ **b** $m \times m \times t$ **c** $p \times p \times t \times t$ **d** $t \times w \times w$

e $m \times q \times m$ **f** $t \times q \times q \times t$ **g** $k \times m \times k \times m$ **h** $t \times y \times y$

i $v \times w \times w \times w$ **j** $p \times d \times d \times d$ **k** $t \times t \times m \times t \times t$ **l** $q \times m \times q$

6 Multiply out each expression.

a $m(m + 3)$ **b** $t(t + 1)$ **c** $k(k + 5)$ **d** $n(n + 2)$

e $t(5 + t)$ **f** $g(2 + g)$ **g** $h(1 + h)$ **h** $d(3 + d)$

i $a(a - 1)$ **j** $c(2 - c)$ **k** $x(1 - x)$ **l** $b(4 - b)$

7 What is the area of each rectangle?

All units are in cm.

a 4, m

b t, $t - 2$

c $m + 8$, m

d $t + 1$, t

(MR) **8** How could Sophia explain the difference to Andrew?

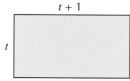

m squared is the same as two m.

No it's not.

Andrew   Sophia

Challenge: Matching multiples

A Show that any three consecutive integers multiplied together always give a multiple of 6.

B Show that the product of any four consecutive integers is always a multiple of 12.

C Is the product of any five consecutive integers always a multiple of 10? Explain why.

Ready to progress?

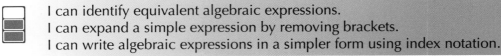

I can simplify algebraic expressions involving multiplication.
I can simplify algebraic expressions by collecting like terms.

I can identify equivalent algebraic expressions.
I can expand a simple expression by removing brackets.
I can write algebraic expressions in a simpler form using index notation.

I can expand an expression involving letters both inside and outside the brackets by removing brackets.

Review questions

1 Write these expressions as simply as possible.

   a $2 \times q$ **b** $m \times 7$ **c** $3 \times y$ **d** $x \times 5$

   e $b \times t$ **f** $m \times p$ **g** $q \times 4$ **h** $m \times 8$

2 Simplify each of these expressions.

   a $n + n$ **b** $g + g + g$ **c** $b + b + b + b$ **d** $e + e + e$

   e $t + t + t + t$ **f** $k + k + k + k$ **g** $n + n + n + n + n$ **h** $g + g + g$

3 Calculate each of these by expanding the bracket.

   a $3(6 + 7)$ **b** $4(5 + 3)$ **c** $6(4 + 2)$

   d $5(3 + 4)$ **e** $5(4 - 3)$ **f** $7(8 - 2)$

4 Write down, as simply as possible, the perimeter of each shape.

   a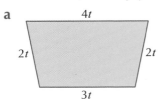   **b**   **c**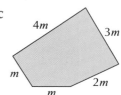

5 What is the area of each rectangle?

   a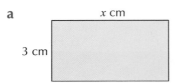   **b**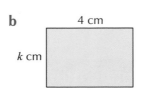

6 Write these expressions as simply as possible.

a $3 \times p \times q$ **b** $4 \times t \times w$ **c** $2 \times m \times t$ **d** $3 \times a \times b$

e $t \times 5 \times m$ **f** $n \times 5 \times p$ **g** $4 \times m \times n$ **h** $b \times 3 \times c$

7 Simplify each expression.

a $3m + 3g + 4m + 2g$ **b** $5t + y + 3t + 3y$ **c** $4g + k + 2g + 2k$

d $6m + 3y - 4m + y$ **e** $5t + 2p - 3t + 3p$ **f** $2p + 4t - p + 3t$

g $5h + 3g - 3h - g$ **h** $8e + 4b - 3e - 3b$ **i** $7m + 3p - 4m - p$

8 The expression in each box is made by adding the expressions in the two boxes it stands on. Copy the diagrams and fill in the missing expressions.

a

b

9 Multiply out each expression.

a $3(t + 4)$ **b** $5(2m + 3)$ **c** $4(3t + 2)$ **d** $3(4x + 1)$

e $5(7 + 3p)$ **f** $4(1 + 3t)$ **g** $5(2 + 3n)$ **h** $2(6 + 4q)$

10 Work out an expression for the area of each rectangle.

Write your answers as simply as possible.

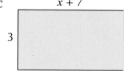

11 Calculate each of these powers.

a 3^2 **b** 4^3 **c** 2^5 **d** 5^2

e 3^3 **f** 2^3 **g** 1^6 **h** 10^3

12 Write each of these as simply as possible.

a $q \times q$ **b** $t + t$ **c** $k \times k \times k$ **d** $m + m + m$

e $w \times w \times w \times w$ **f** $k + k + k + k$ **g** $d + d$ **h** $g \times g \times g$

i $t + t + t + t$ **j** $y \times y \times y \times y \times y$ **k** $x + x + x$ **l** $n \times n \times n \times n$

13 Multiply out each of these expressions.

a $t(t + 2)$ **b** $m(m + 3)$ **c** $h(h + 4)$ **d** $x(x + 1)$

e $p(4 + p)$ **f** $k(4 + k)$ **g** $y(2 + y)$ **h** $e(5 + e)$

i $b(b - 1)$ **j** $a(3 - a)$ **k** $c(1 - c)$ **l** $m(5 - m)$

Mathematical reasoning
Strawberries

Strawberries are a favourite with many people. You can eat them at breakfast, with afternoon tea and for a lovely pudding.

Strawberry Facts

Here are some facts about strawberries.

◎ Each strawberry carries about 200 seeds on the outside.

◎ Six strawberries a day will give you all the vitamin C that you need.

◎ Madam Tallein, a member of the court of Napoleon, used to take a bath in fresh strawberry juice using 22 pounds of strawberries for each bath.

1 a In a punnet of strawberries there are 30 strawberries. How many seeds are in that punnet?

b How many seeds are there on x strawberries?

2 a How many strawberries a day will a family of four need to ensure they have all the vitamin C that they need?

b How many strawberries are needed each day for a family of k people to have sufficient vitamin C?

3 1 kg is equivalent to 2.2 pounds.

a How many kg are equivalent to 22 pounds?

b When Madam Tallein had three baths a week, how many kilograms of strawberries would she need each week to make the strawberry juice?

c When Madam Tallein had B baths a week, how many kilograms of strawberries would she need each week to make the strawberry juice?

Growing strawberries

To grow strawberries, start with young plants and plant them in rows so that each strawberry plant is 40 cm away from each other.

Each row should be 120 cm away from the next row.

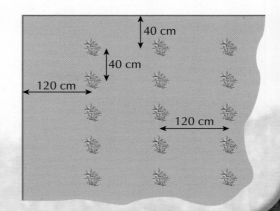

4 How many strawberry plants will I need if I want to:

a plant four rows with 20 in each row

b plant four rows with R in each row

c plant P rows with 20 in each row

d plant P rows with R in each row?

5 I plant three rows with four plants in every row. The plot size I need is shown in the diagram.

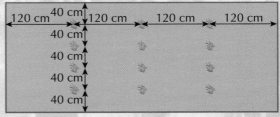

a How wide must the plot when there are:

i 8 rows   **ii** x rows?

b How long does the plot need to be when, in each row, there are:

i 10 plants   **ii** y plants?

6 **a** How many rows and columns could you get out of a plot that is 1200 cm wide and 800 cm deep?

b How many strawberry plants would you need for a plot 1200 cm by 800 cm?

c Explain how you would find how many plants you need for a plot measuring:

i 960 cm by 480 cm   **ii** x cm by y cm.

11

Congruence and scaling

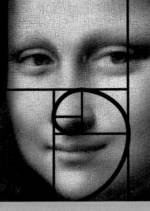

This chapter is going to show you:

- how to recognise congruent shapes
- how to use shape and ratio
- how to use scale diagrams.

You should already know:

- how to work out perimeters and areas of squares and rectangles
- how to use ratio.

About this chapter

In a golden rectangle, the side lengths are in the ratio 1 : φ.

φ is the Greek letter phi and is approximately equal to 1.618.

This rectangle is special, because if you cut a square from one end of it, you will be left with a smaller shape that is another golden rectangle, with sides that are in the same ratio as the one you started with.

The golden rectangle has been described as one of the most visually pleasing rectangular shapes. Many artists and architects have used golden rectangles in their work.

For example, Leonardo da Vinci's Mona Lisa and Salvador Dali's The Sacrament of the Last Supper are golden rectangles and the exterior of the Parthenon on the Acropolis in Athens is based on the golden rectangle.

The flag of Togo was designed to approximate a golden rectangle. The same shape can also be seen in flower seed heads, and in the shells of sea creatures.

11.1 Congruent shapes

Learning objective

* To recognise congruent shapes

All the triangles on the grid below are reflections, rotations or translations of triangle A.

What do you notice about them?

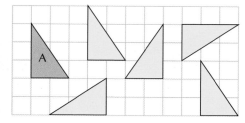

All the triangles are exactly the same shape and size as triangle A.

Two shapes that are exactly the same shape and size are said to be **congruent**. One shape may be turned over or turned round compared with the other shape.

Reflections, rotations and translations all produce images that are congruent to the original shape.

Example 1

Which two shapes are congruent?

A   B   C   D

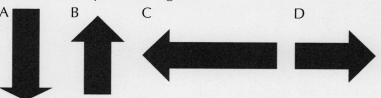

  Shapes B and D are exactly the same shape and size, so they are congruent.

  Use tracing paper to check that the two shapes are congruent.

Example 2

Which isosceles triangle is not congruent to the others?

  Triangle D.

  All the triangles are 2 squares long on the shortest side.

  Triangles A, B and C are 3 squares high, but triangle D is 4 squares high.

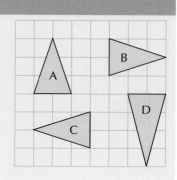

Exercise 11A

1 For each pair of shapes below, state whether or not they are congruent.
   Use tracing paper to help if you are not sure.

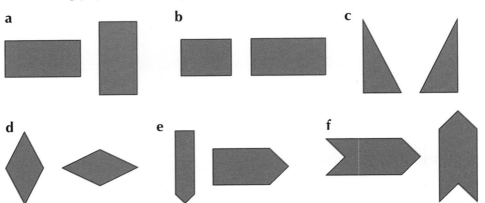

2 Look carefully at the triangles in this diagram.

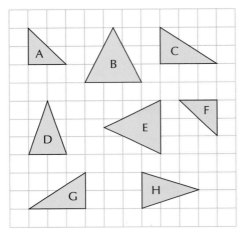

Now copy and complete these sentences about the triangles.

a Triangle A is congruent to triangle … .
b Triangle B is congruent to triangle … .
c Triangle C is congruent to triangle … .
d Triangle D is congruent to triangle … .

3 Match up the pairs of congruent T-shapes.

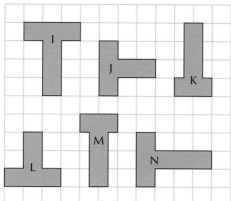

4 Which shape is the odd one out?

Give a reason for your answer.

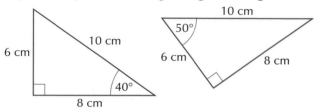

a　　　　　　　　**b**　　**c**　　　　**d**

MR **5** Explain why these two right-angled triangles are congruent.

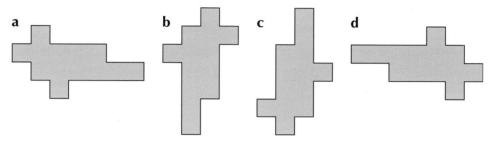

6 The six arrow-shapes are all congruent.

Use each word once to complete the sentences.

translation　　　reflection　　　rotation

a Arrow B is a .... of arrow A.

b Arrow D is a .... of arrow C.

c Arrow F is a .... of arrow E.

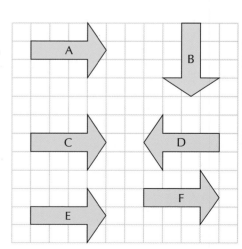

7 On a copy of the grid, draw six more right-angled triangles that are congruent to triangle X.

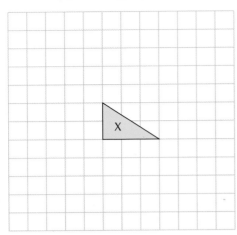

Activity: Making shapes from congruent triangles

Two congruent right-angled triangles are placed together with two of their equal sides touching to make another shape, as shown in the diagram.

 +   make an isosceles triangle: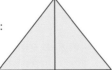

A How many different shapes can you make? To help, cut out the triangles from a piece of card.

B Repeat the activity using two congruent isosceles triangles.

 +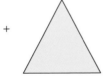

C Repeat the activity using two congruent equilateral triangles.

 +

11.2 Shape and ratio

Learning objective

- To use ratio to compare lengths and areas of 2D shapes

Ratio can be used to compare lengths and areas of 2-D shapes, as these examples show.

Example 3

Work out the ratio of the length of the line AB to the length of the line CD.

A —————B   C ————————————————D
12 mm   4.8 cm

First convert the measurements to the same units and then simplify the ratio.

Always use the smaller unit, which here is millimetres (mm).

    4.8 cm = 48 mm

Then the ratio is 12 mm : 48 mm = 1 : 4.

 Hint   Remember that ratios do not have units.

Example 4

Work out the ratio of:

a the perimeter of rectangle A to the perimeter of rectangle B, giving the answer in its simplest form

b the area of rectangle A to the area of rectangle B, giving the answer in its simplest form.

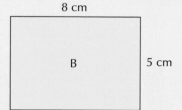

  a The ratio of the perimeters is 14 cm : 26 cm   Divide by 2.

     = 7 : 13

  b The ratio of the areas is 12 cm^2 : 40 cm^2 Divide by 4.

     = 3 : 10

1 For each pair of lines, work out the ratio of the length of the line AB to the length of the line CD.

Remember to simplify the ratio where possible.

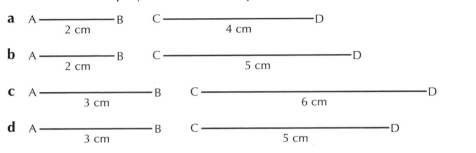

a A————B 2 cm   C————————D 4 cm

b A————B 2 cm   C—————————————D 5 cm

c A——————B 3 cm   C———————————————D 6 cm

d A——————B 3 cm   C———————————D 5 cm

e A————————B 4 cm   C———————————————D 6 cm

2 Write down each ratio in its simplest form.

a 15 mm : 20 mm   **b** 18 cm : 30 cm   **c** 8 m : 12 m

d 40 mm : 5 cm   **e** 25 cm : 2 m   **f** 50 cm : 1.5 m

3 For each pair of rectangles, work out the ratio of:

i the perimeter of rectangle A to the perimeter of rectangle B, giving your answer in its simplest form

ii the area of rectangle A to the area of rectangle B, giving your answer in its simplest form.

a A: 3 cm × 2 cm   B: 4 cm × 3 cm

b A: 4 cm × 2 cm   B: 5 cm × 3 cm

c A: 5 cm × 2 cm   B: 10 cm × 3 cm

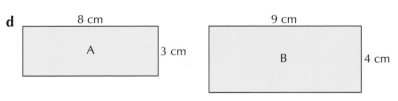

d A: 8 cm × 3 cm   B: 9 cm × 4 cm

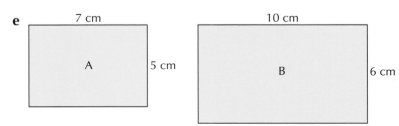

e A: 7 cm × 5 cm   B: 10 cm × 6 cm

4 For each pair of squares, work out the ratio of:

  i the perimeter of square X to the perimeter of square Y, giving your answer in its simplest form

  ii the area of square X to the area of square Y, giving your answer in its simplest form.

a
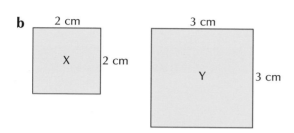

b
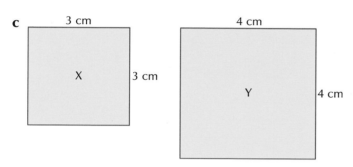

c

5 The dimensions of lawn A and lawn B are given on the diagrams.

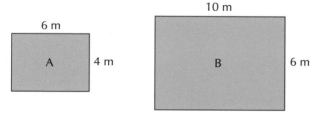

  a Calculate the area of lawn A.

  b Calculate the area of lawn B.

  c Write down the ratio of the length of lawn A to the length of lawn B, giving your answer in its simplest form.

  d Write down the ratio of the area of lawn A to the area of lawn B, giving your answer in its simplest form.

  e Write down the area of lawn A as a fraction of the area of lawn B, giving your answer in its simplest form.

(PS) 6 a Work out the ratio of the area of the blue square to the area of the red surround, giving your answer in its simplest form.

b Write down the area of the blue square as a fraction of the area of the red surround.

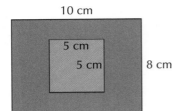

Investigation: Growing rectangles

The dimensions of three rectangles A, B and C are given in the diagram.

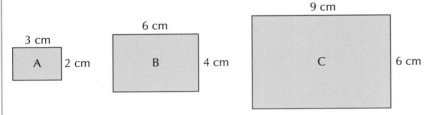

The ratio of the length of A to the length of B to the length of C is 3 cm : 6 cm : 9 cm = 1 : 2 : 3.

Work out each ratio in the same way, giving your answers in their simplest form.

A The width of A to the width of B to the width of C.

B The perimeter of A to the perimeter of B to the perimeter of C.

C The area of A to the area of B to the area of C.

D The dimensions of the rectangles form a pattern.

Rectangle D is added to the pattern.

Write down the dimensions of rectangle D.

E Write down each ratio.

a The length of A to the length of B to the length of C to the length of D.

b The width of A to the width of B to the width of C to the width of D.

c The perimeter of A to the perimeter of B to the perimeter of C to the perimeter of D.

d The area of A to the area of B to the area of C to the area of D.

11.3 Scale diagrams

Learning objective

● To understand and use scale diagrams

A **scale diagram** is a smaller drawing of an actual object. A **scale** must always be clearly given by the side of or below the scale diagram.

This is part of an architect's blue-print. It is a scale diagram for an extension to a building.

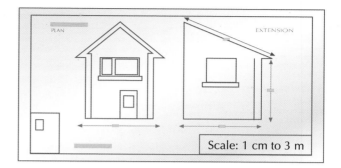

Example 5

These lines are drawn using a scale of 1 cm to 5 m.

Work out the length each line represents.

a ──────
b ─────────
c ───────────

  Measure each line.

  a The length of the line is 2 cm, so it represents 2 × 5 = 10 m.

  b The length of the line is 3.5 cm, so it represents 3.5 × 5 = 17.5 m.

  c The length of the line is 6.2 cm, so it represents 6.2 × 5 = 31 m.

Example 6

This is a scale diagram of Rebecca's room.

Work out:

a the actual length of the room

b the actual width of the room

c the actual width of the window.

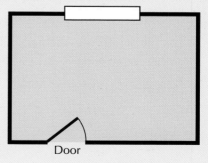

Door

Scale: 1 cm to 1 m

  Measure each length on the scale diagram.

  a On the scale diagram, the length of the room is 5 cm, so the actual length of the room is 5 m.

  b On the scale diagram, the width of the room is 3.5 cm, so the actual width of the room is 3.5 m.

  c On the scale diagram, the width of the window is 2 cm, so the actual width of the window is 2 m.

1. The lines shown are drawn using a scale of 1 cm to 10 m. Write down the length each line represents.

    a ─────────

    b ─────────────

    c ───────────────

    d ──────────────────

    e ────────────────────────

2. This is a scale diagram of a school hall.

    a   Write down the actual length of the hall.

    b   Write down the actual width of the hall.

    c   Work out the actual area of the hall.

    Scale: 1 cm to 4 m

3. Copy and complete this table for a scale diagram where the scale is 1 cm to 5 m.

	Length on scale diagram	Actual length
a	2 cm	
b	5 cm	
c	6.5 cm	
d		15 m
e		40 m
f		52.5 m

4. This is a scale diagram of a kite.

    a   Measure the lengths of AC and BD.

    b   Write down the lengths AC and BD on the actual kite.

    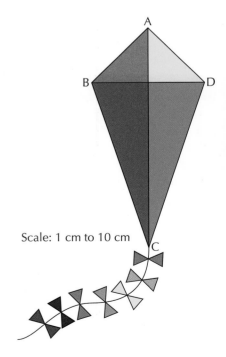

    Scale: 1 cm to 10 cm

5 The diagram shows Ryan's scale diagram for his mathematics classroom.

Nathan notices that Ryan has not put a scale on the diagram. He knows that the width of the classroom is 8 m.

a What scale has Ryan used?

b What is the actual length of the classroom?

c What is the actual area of the classroom?

 6 The diagram shows the actual measurements of a football pitch.

It is not drawn to scale.

Use the measurements on the diagram to make a scale diagram of the pitch on centimetre-squared paper.

Use a scale of 1 cm to 10 m.

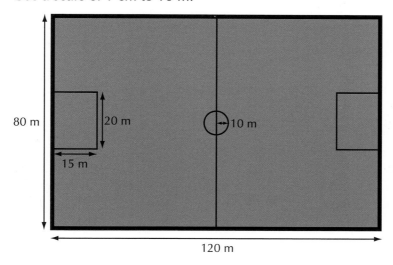

Activity: Scale diagram of a classroom

You will need a metre ruler or a tape measure for this activity.

On centimetre-squared paper, draw a scale diagram of your classroom, including the desks and any other furniture in the room.

Use a scale of 2 cm to 1 m.

Ready to progress?

I can recognise congruent shapes.
I can solve shape problems using ratio.
I can use scale diagrams.

Review questions

1 Copy and complete these sentences.

   a Shape A is congruent to shape....

   b Shape B is congruent to shape....

   c Shape C is congruent to shape....

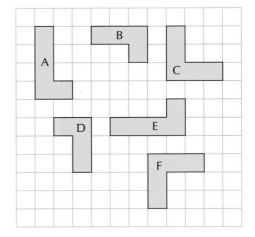

 2 The diagram shows how four congruent trapezia can make a parallelogram.

   Four congruent trapezia can also make a bigger trapezium.

   On dotty triangle paper, draw a diagram to show how four trapezia can make a bigger trapezium.

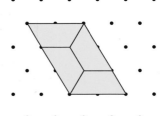

3 For each of the following, measure the length of the lines to work out the ratio of the length of the line PQ to the length of the line RS.

   Remember to simplify the ratio where possible.

   a P————Q   R—————S

   b P—————Q   R——————S

   c P———Q   R——————S

4   For each pair of regular polygons work out the ratio of the perimeter of polygon A to the perimeter of polygon B, giving your answer in its simplest form.

a regular pentagons

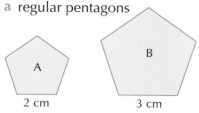

2 cm   3 cm

b regular hexagons

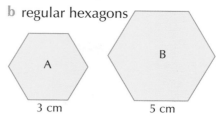

3 cm   5 cm

c regular octagons

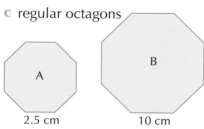

2.5 cm   10 cm

5   These pictures are drawn using a scale of 1 cm to 40 cm.

a   Work out the actual height and width of the door.
b   Work out the length and height of the actual car, giving your answers in metres.

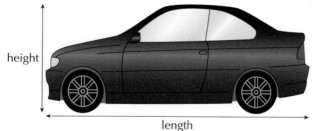

height

length

c   Work out the length of the actual canoe, giving your answers in metres.

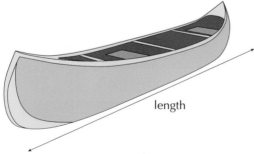

length

 6 a   Work out the ratio of the area of the red squares to the area of the whole square, giving your answer in its simplest form.

b   Work out the ratio of the area of the red squares to the area of the blue rectangles, giving your answer in its simplest form.

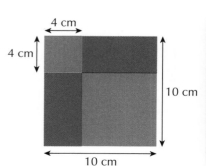

4 cm

4 cm

10 cm

10 cm

Financial skills
Carpeting a bungalow

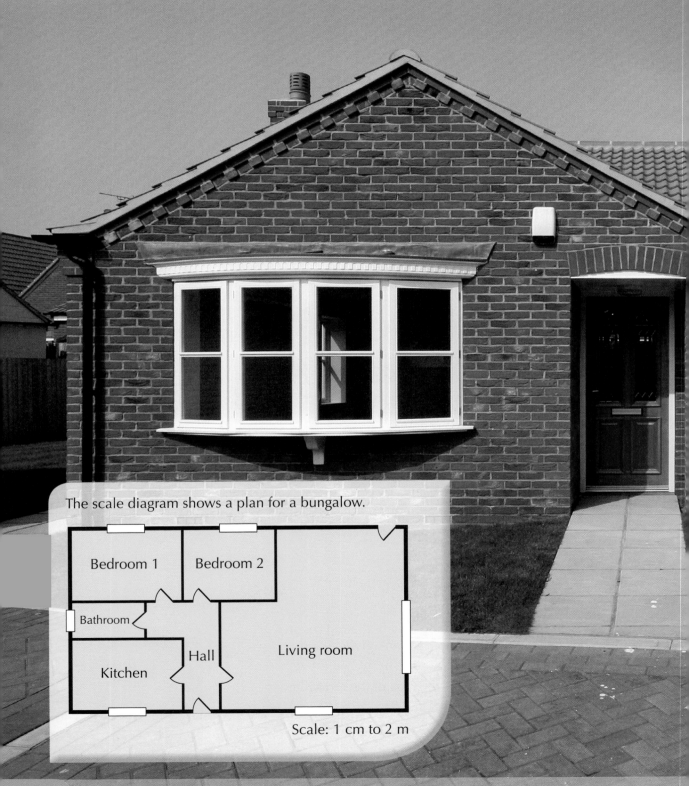

The scale diagram shows a plan for a bungalow.

Bedroom 1

Bedroom 2

Bathroom

Kitchen

Hall

Living room

Scale: 1 cm to 2 m

1 Work out the actual dimensions of:

   a the kitchen   **b** the bathroom   **c** bedroom 1   **d** bedroom 2.

2 Write down the actual area of:

   a the kitchen   **b** the bathroom   **c** bedroom 1   **d** bedroom 2.

3 Bedroom 1 and bedroom 2 are to have the same carpets.

   Work out the total cost of the carpeting when 1 m^2 of carpet costs £4.49 and 1 m^2 of underlay costs £2.26.

4 Work out:

   a the actual perimeter of the living room

   b the actual area of the living room.

5 The living room is to have a different carpet.

   Work out the total cost of the carpeting when 1 m^2 of carpet costs £5.49 and 1 m^2 of underlay costs £3.53.

12

Fractions and decimals

This chapter is going to show you:

- how to multiply a fraction by an integer
- how to divide with unit fractions and integers
- how to multiply by powers of ten
- how to divide by powers of ten.

You should already know:

- the relationship between mixed numbers and improper fractions
- how to add two simple fractions
- how to subtract two simple fractions.

About this chapter

Fractions have been written in different ways during human history. Nowadays we use two different ways of writing fractional numbers – either as one whole number over another whole number, or using a decimal point.

Why do we need to know both? Decimals are often more convenient but there are some simple fractions that are not easy to write as decimals. For example:

- $\frac{2}{3}$ as a decimal is 0 666 666 6 ...
- $\frac{1}{7}$ as a decimal is 0.142 857 142 ...

Because of this it is useful to be able to write and calculate with fractional numbers in both ways.

In this chapter you will be learning to do calculations with fractions and with decimals. In particular you will be looking at multiplication and division.

12.1 Adding and subtracting fractions

Learning objective

- To add and subtract fractions and mixed numbers

You should already know how to add and subtract fractions. The examples below will remind you.

You often have to find equivalent fractions with the same denominator. When you do this, find the lowest common multiple of the two denominators you have. This can be the new denominator for both fractions.

Example 1

Add these numbers. **a** $\frac{3}{4} + \frac{5}{8}$ **b** $\frac{2}{3} - \frac{1}{4}$

a Change $\frac{3}{4}$ to eighths.

$$\frac{3}{4} + \frac{5}{8}$$

$$= \frac{6}{8} + \frac{5}{8} = \frac{11}{8}$$

$$= 1\frac{3}{8}$$

b $\frac{2}{3} - \frac{1}{4}$

$$= \frac{8}{12} - \frac{3}{12}$$

$$= \frac{5}{12}$$

The fractions must have the same denominator.

$$\frac{3}{4} = \frac{3 \times 2}{4 \times 2} = \frac{6}{8}$$

Add the numerators ($6 + 5 = 11$). The denominator does not change.

Change the improper fraction to a mixed number.
$11 \div 8 = 1$ remainder 3

Change to twelfths because 3 and 4 are factors of 12.

Multiply 2 and 3 by 4; multiply 1 and 4 by 3.

$8 - 3 = 5$. The denominator does not change.

Example 2

Calculate: **a** $1\frac{1}{2} + \frac{3}{4}$ **b** $1\frac{1}{2} - \frac{7}{8}$

a $1\frac{1}{2} + \frac{3}{4}$

$$= \frac{3}{2} + \frac{3}{4}$$

$$= \frac{6}{4} + \frac{3}{4}$$

$$= \frac{9}{4}$$

$$= 2\frac{1}{4}$$

b $1\frac{1}{2} - \frac{7}{8}$

$$= \frac{3}{2} - \frac{7}{8}$$

$$= \frac{12}{8} - \frac{7}{8}$$

$$= \frac{5}{8}$$

Change the mixed number to an improper fraction.

Change $\frac{3}{2}$ to $\frac{6}{4}$ by multiplying 3 and 2 by 2.

$6 + 3 = 9$. The denominator does not change.

Change to a mixed number.

Change $1\frac{1}{2}$ to an improper fraction.

Change $\frac{3}{2}$ to eighths: $\frac{3}{2} = \frac{3 \times 4}{2 \times 4} = \frac{12}{8}$

Exercise 12A

1 Copy and complete these sets of equivalent fractions.

   a $\dfrac{2}{3} = \dfrac{\square}{12}$
   b $\dfrac{3}{4} = \dfrac{\square}{12}$
   c $\dfrac{3}{2} = \dfrac{\square}{8}$
   d $\dfrac{5}{3} = \dfrac{\square}{9}$
   e $\dfrac{7}{4} = \dfrac{\square}{12}$

2 Write each number as an improper fraction.

   a $1\frac{1}{4}$
   b $1\frac{5}{8}$
   c $3\frac{1}{2}$
   d $2\frac{4}{5}$
   e $2\frac{2}{3}$
   f $3\frac{1}{4}$

3 Write each fraction as a mixed number as simply as possible.

   a $\dfrac{8}{5}$
   b $\dfrac{6}{4}$
   c $\dfrac{11}{3}$
   d $\dfrac{10}{6}$
   e $\dfrac{20}{8}$
   f $\dfrac{13}{4}$

4 Add these fractions. Write the answers as simply as possible. Write the answers as mixed numbers if necessary.

   a $\dfrac{1}{8} + \dfrac{3}{8}$
   b $\dfrac{1}{3} + \dfrac{1}{3}$
   c $\dfrac{2}{5} + \dfrac{2}{5}$
   d $\dfrac{5}{8} + \dfrac{1}{8}$

   e $\dfrac{1}{6} + \dfrac{5}{6}$
   f $\dfrac{3}{4} + \dfrac{3}{4}$
   g $\dfrac{5}{8} + \dfrac{3}{8}$
   h $\dfrac{7}{8} + \dfrac{5}{8}$

5 Subtract these fractions. Write the answers as simply as possible.

   a $\dfrac{5}{8} - \dfrac{3}{8}$
   b $\dfrac{5}{6} - \dfrac{1}{6}$
   c $\dfrac{7}{8} - \dfrac{1}{8}$
   d $\dfrac{7}{10} - \dfrac{3}{10}$

   e $\dfrac{7}{12} - \dfrac{1}{12}$
   f $\dfrac{11}{12} - \dfrac{5}{12}$
   g $\dfrac{9}{16} - \dfrac{7}{16}$
   h $\dfrac{15}{16} - \dfrac{3}{16}$

6 Add these fractions.

   a $\dfrac{1}{2} + \dfrac{1}{4}$
   b $\dfrac{1}{2} + \dfrac{1}{8}$
   c $\dfrac{3}{8} + \dfrac{1}{2}$
   d $\dfrac{1}{2} + \dfrac{1}{6}$

   e $\dfrac{1}{4} + \dfrac{1}{8}$
   f $\dfrac{3}{8} + \dfrac{1}{4}$
   g $\dfrac{3}{4} + \dfrac{1}{8}$
   h $\dfrac{5}{8} + \dfrac{1}{4}$

7 Subtract these fractions.

   a $\dfrac{1}{2} - \dfrac{1}{4}$
   b $\dfrac{1}{2} - \dfrac{1}{8}$
   c $\dfrac{3}{8} - \dfrac{1}{4}$
   d $\dfrac{1}{2} - \dfrac{1}{6}$

   e $\dfrac{3}{4} - \dfrac{1}{8}$
   f $\dfrac{5}{8} - \dfrac{1}{4}$
   g $\dfrac{3}{4} - \dfrac{3}{8}$
   h $\dfrac{3}{4} - \dfrac{5}{12}$

8 Add these numbers. Give your answers as simply as possible, as mixed numbers.

   a $1\frac{1}{4} + \frac{1}{2}$
   b $\frac{3}{4} + 1\frac{1}{2}$
   c $1\frac{1}{6} + \frac{2}{3}$
   d $1\frac{3}{10} + \frac{2}{5}$

   e $\frac{1}{2} + 1\frac{3}{4}$
   f $\frac{1}{4} + 2\frac{1}{2}$
   g $2\frac{1}{4} + 1\frac{1}{2}$
   h $1\frac{1}{2} + 1\frac{3}{4}$

9 Subtract these numbers. Give your answers as simply as possible, as mixed numbers.

   a $1\frac{1}{4} - \frac{1}{2}$
   b $1\frac{3}{4} - \frac{1}{2}$
   c $1\frac{1}{6} - \frac{2}{3}$
   d $1\frac{3}{5} - \frac{7}{10}$

   e $2\frac{1}{2} - \frac{3}{4}$
   f $1\frac{3}{8} - \frac{1}{2}$
   g $2\frac{1}{4} - 1\frac{1}{2}$
   h $2\frac{1}{2} - 1\frac{3}{4}$

10 Work these out.

   a $\dfrac{1}{2} + \dfrac{1}{3}$
   b $\dfrac{2}{3} - \dfrac{1}{2}$
   c $\dfrac{3}{4} + \dfrac{1}{6}$
   d $\dfrac{5}{6} - \dfrac{1}{3}$

11   **a**   Write $1\frac{1}{2}$ in sixths.   **b**   Write $1\frac{2}{3}$ in sixths.   **c**   Work out $1\frac{1}{2} + 1\frac{2}{3}$.

12   Work these out.

    a $2\frac{1}{2} + 1\frac{1}{3}$ **b** $1\frac{1}{2} + 2\frac{1}{3}$ **c** $2\frac{2}{3} - 1\frac{1}{2}$ **d** $3\frac{1}{2} - 1\frac{2}{3}$

13   **a**   Write $1\frac{1}{4}$ in twelfths.   **b**   Write $1\frac{5}{6}$ in twelfths.   **c**   Work out $1\frac{1}{4} + 1\frac{5}{6}$.

14   Work these out.

    a $1\frac{1}{4} + 1\frac{1}{6}$ **b** $1\frac{5}{6} + 2\frac{1}{4}$ **c** $2\frac{3}{4} - 1\frac{1}{6}$ **d** $3\frac{1}{6} - 1\frac{3}{4}$

(PS)   **15**   Find the missing number in each addition.

    a $1\frac{1}{4} + \ldots = 3$ **b** $1\frac{3}{4} + \ldots = 2\frac{1}{4}$ **c** $1\frac{1}{2} + \ldots = 3\frac{3}{4}$ **d** $2\frac{1}{2} + \ldots = 4\frac{1}{4}$

Challenge: Magic square

This is a fractional magic square.

$\frac{2}{15}$		
$\frac{7}{15}$	$\frac{5}{15}$	$\frac{3}{15}$

Every row, column and diagonal has the same total, called the magic number.

A   What is the magic number?

B   Copy the magic square and fill in the missing fractions.

C   Redraw the magic square but write each fraction in its simplest possible terms.

12.2 Multiplying fractions and integers

Learning objective

- To multiply a fraction or a mixed number by an integer

Look at these three calculations.

$\frac{1}{4}$ of 6 $\frac{1}{4} \times 6$ $6 \times \frac{1}{4}$

They all have the same answer.

$\frac{1}{4}$ of $6 = 6 \div 4 = 1\frac{2}{4} = 1\frac{1}{2}$

$\frac{1}{4} \times 6 = \frac{1}{4} + \frac{1}{4} + \frac{1}{4} + \frac{1}{4} + \frac{1}{4} + \frac{1}{4} = \frac{6}{4} = 1\frac{2}{4} = 1\frac{1}{2}$

$6 \times \frac{1}{4}$ is the same as $\frac{1}{4} \times 6$ because when you multiply two numbers, the order does not matter.

Example 3

Work out: **a** $\frac{1}{2} \times 8$ **b** $\frac{1}{3} \times 5$ **c** $7 \times \frac{3}{5}$.

a $\frac{1}{2} \times 8 = \frac{1 \times 8}{2}$ Multiply the numerator by 8. Keep the same denominator.

   $= \frac{8}{2} = 4$ $8 \div 2 = 4$. There is no remainder.

b $\frac{1}{3} \times 5 = \frac{1 \times 5}{3}$ Multiply the numerator by 5.

   $= \frac{5}{3} = 1\frac{2}{3}$ $5 \div 3 = 1$ remainder 2

c $7 \times \frac{3}{5} = \frac{21}{5}$ $7 \times 3 = 21$ and keep the same denominator.

   $= 4\frac{1}{5}$ $21 \div 5 = 4$ remainder 1

The next example shows how to multiply a mixed number by an integer.

Example 4

Work out $2\frac{1}{2} \times 4$.

$2\frac{1}{2} \times 4 = \frac{5}{2} \times 4$ Write $2\frac{1}{2}$ as a mixed number.

$= \frac{5 \times 4}{2}$

$= \frac{20}{2} = 10$

Exercise 12B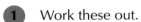

1 Work these out.

   a $\frac{1}{3}$ of 9   **b** $\frac{1}{3}$ of 15   **c** $\frac{1}{3}$ of 24   **d** $\frac{1}{4}$ of 20

2 Work these out.

   a $\frac{1}{2} \times 12$ **b** $\frac{1}{2} \times 20$ **c** $\frac{1}{3} \times 12$ **d** $\frac{1}{4} \times 12$

   e $18 \times \frac{1}{3}$ **f** $30 \times \frac{1}{5}$ **g** $24 \times \frac{1}{8}$ **h** $28 \times \frac{1}{4}$

3 Work these out.

   a $\frac{1}{2} \times 5$ **b** $\frac{1}{2} \times 11$ **c** $\frac{1}{3} \times 14$ **d** $\frac{1}{4} \times 17$

   e $17 \times \frac{1}{3}$ **f** $21 \times \frac{1}{5}$ **g** $5 \times \frac{1}{8}$ **h** $33 \times \frac{1}{4}$

4 Work these out.

   a $\frac{1}{3}$ of 17   **b** $\frac{1}{5}$ of 14   **c** $\frac{1}{3}$ of 16   **d** $\frac{1}{4}$ of 19

   e $\frac{1}{6}$ of 5   **f** $\frac{1}{8}$ of 21   **g** $\frac{1}{8}$ of 3   **h** $\frac{1}{5}$ of 42

5 Write your answers to these as simply as possible.

   a $\frac{1}{4} \times 6$ **b** $\frac{1}{4} \times 14$ **c** $\frac{1}{4} \times 22$ **d** $\frac{1}{6} \times 15$

   e $10 \times \frac{1}{6}$ **f** $25 \times \frac{1}{10}$ **g** $4 \times \frac{1}{8}$ **h** $4 \times \frac{1}{6}$

6 Work these out.

a $\frac{2}{3} \times 6$ b $\frac{2}{3} \times 12$ c $\frac{3}{4} \times 4$ d $\frac{3}{4} \times 8$

e $10 \times \frac{4}{5}$ f $12 \times \frac{5}{6}$ g $15 \times \frac{2}{5}$ h $9 \times \frac{2}{3}$

7 Work these out.

a $\frac{2}{3} \times 2$ b $\frac{2}{3} \times 5$ c $\frac{3}{4} \times 3$ d $\frac{3}{4} \times 5$

e $3 \times \frac{4}{5}$ f $7 \times \frac{3}{5}$ g $5 \times \frac{5}{8}$ h $7 \times \frac{7}{8}$

8 Write the answers to these as simply as possible.

a $\frac{3}{4} \times 2$ b $\frac{3}{4} \times 6$ c $\frac{3}{8} \times 4$ d $\frac{3}{8} \times 2$

e $4 \times \frac{5}{6}$ f $12 \times \frac{3}{8}$ g $4 \times \frac{3}{8}$ h $8 \times \frac{5}{12}$

9 Work these out.

a $1\frac{1}{4} \times 4$ b $1\frac{3}{4} \times 4$ c $2\frac{1}{2} \times 4$ d $1\frac{2}{3} \times 3$

e $6 \times 1\frac{1}{3}$ f $8 \times 2\frac{1}{4}$ g $12 \times 1\frac{1}{4}$ h $6 \times 2\frac{1}{3}$

(PS) **10** Work out the area of each rectangle.

a
4 cm
$2\frac{1}{4}$ cm

b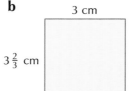
3 cm
$3\frac{2}{3}$ cm

c
4 cm
$4\frac{3}{4}$ cm

11 Work these out.

a $1\frac{1}{2} \times 3$ b $1\frac{3}{4} \times 5$ c $1\frac{2}{5} \times 4$ d $2\frac{2}{3} \times 4$

e $2\frac{1}{4} \times 5$ f $3\frac{2}{3} \times 5$ g $2\frac{3}{4} \times 3$ h $1\frac{1}{3} \times 8$

12 $33\frac{1}{3}\% = \frac{1}{3}$

Use this fact to work out:

a $33\frac{1}{3}\%$ of 12   b $33\frac{1}{3}\%$ of 10   c $33\frac{1}{3}\%$ of 17.

13 $66\frac{2}{3}\% = \frac{2}{3}$

Use this fact to work out:

a $66\frac{2}{3}\%$ of 18   b $66\frac{2}{3}\%$ of 13   c $66\frac{2}{3}\%$ of 23.

Challenge: Multiplication table

Copy and complete this multiplication table.

×	2	3	4	5
$1\frac{1}{2}$				
$1\frac{3}{4}$		$5\frac{1}{4}$		
$1\frac{2}{3}$				
$2\frac{1}{3}$			$9\frac{1}{3}$	

12.3 Dividing with integers and fractions

Learning objectives

- To divide a unit fraction by an integer
- To divide an integer by a unit fraction

Key word

unit fraction

In this section you will learn about division with a **unit fraction** and an integer. A unit fraction is a fraction with a numerator of 1, such as $\frac{1}{2}$ or $\frac{1}{5}$ or $\frac{1}{12}$.

Look at these two examples of division.

- $\frac{1}{2} \div 3$
- $3 \div \frac{1}{2}$

They are *not* the same.

$\frac{1}{2} \div 3$ means divide a half into three equal parts.

The answer is $\frac{1}{6}$ because $\frac{1}{6} \times 3 = \frac{3}{6} = \frac{1}{2}$.

$3 \div \frac{1}{2}$ means work out how many halves make 3.

The answer is 6 because $\frac{1}{2} \times 6 = \frac{6}{2} = 3$.

$$\frac{1}{2} \div 3 = \frac{1}{6} \qquad 3 \div \frac{1}{2} = 6$$

Example 5

Work out: **a** $\frac{1}{4} \div 2$ **b** $4 \div \frac{1}{5}$.

a $\frac{1}{4} \div 2 = \frac{1}{4 \times 2} = \frac{1}{8}$ Check that $2 \times \frac{1}{8} = \frac{2}{8} = \frac{1}{4}$.

b $4 \div \frac{1}{5} = 4 \times 5 = 20$ Check that $\frac{1}{5} \times 20 = \frac{20}{5} = 4$.

Exercise 12C

1. Work these out.

   a $\frac{1}{2} \div 2$ **b** $\frac{1}{2} \div 3$ **c** $\frac{1}{2} \div 5$ **d** $\frac{1}{2} \div 6$

2. Work these out.

   a $\frac{1}{3} \div 2$ **b** $\frac{1}{3} \div 4$ **c** $\frac{1}{3} \div 8$ **d** $\frac{1}{3} \div 10$

3. Work these out.

   a $\frac{1}{4} \div 3$ **b** $\frac{1}{5} \div 2$ **c** $\frac{1}{8} \div 3$ **d** $\frac{1}{4} \div 6$

4. Work these out.

   a $1 \div \frac{1}{2}$ **b** $2 \div \frac{1}{2}$ **c** $5 \div \frac{1}{2}$ **d** $9 \div \frac{1}{2}$

5. Work these out.

   a $1 \div \frac{1}{3}$ **b** $2 \div \frac{1}{3}$ **c** $4 \div \frac{1}{3}$ **d** $6 \div \frac{1}{3}$

6. Work these out.

   a $1 \div \frac{1}{4}$ **b** $4 \div \frac{1}{4}$ **c** $5 \div \frac{1}{4}$ **d** $10 \div \frac{1}{4}$

7 Work these out.

a $6 \div \frac{1}{4}$ b $\frac{1}{4} \div 6$ c $9 \div \frac{1}{5}$ d $\frac{1}{5} \div 9$

e $\frac{1}{3} \div 12$ f $12 \div \frac{1}{3}$ g $\frac{1}{10} \div 15$ h $15 \div \frac{1}{10}$

(FS) 8 50p is £$\frac{1}{2}$.

a Copy and complete:

The number of 50p coins that make £7 is $7 \div \frac{1}{2} = \ldots$

b Work out how many 50p coins make:

i £3   ii £9   iii £15   iv £24   v £51.

(FS) 9 20p is £$\frac{1}{5}$.

a Copy and complete:

The number of 20p coins that make £4 is $4 \div \frac{1}{5} = \ldots$

b Work out how many 20p coins make:

i £3   ii £6   iii £9   iv £15   v £21.

(PS) 10 a The perimeter of a regular hexagon is $\frac{1}{2}$ metre.

Work out the length of each side.

b Each side of a different regular polygon is $\frac{1}{3}$ m long.

The perimeter of this polygon is 4 m.

How many sides does it have?

Challenge: Cycle race

Cycle races take place in a velodrome.

Older tracks have a length of $\frac{1}{3}$ km but modern tracks are $\frac{1}{4}$ km.

A Copy and complete this sentence.

The number of laps in a 5 km race on an older track is $5 \div \frac{1}{3} = \ldots$.

B Copy and complete this sentence.

The number of laps in a 5 km race on a modern track is $5 \div \ldots = \ldots$.

C In 1996, Chris Boardman set a one hour cycling record of 56 km.

Copy and complete this sentence.

On an older track, 56 km is $56 \div \ldots = \ldots$ laps.

12.4 Multiplication with powers of ten

Learning objective

• To multiply by a power of ten mentally

These numbers are all **powers of 10**: 1000, 100, 10, 0.1, 0.01.

You should be able to multiply any number by a power of 10 easily, without using a calculator.

Here are some examples, showing how the digits stay the same

$3.4 \times 10 = 34$ $3.4 \times 100 = 340$ $3.4 \times 0.1 = 0.34$

H T U . t h   H T U . t h   H T U . t h

      3 . 4   3 . 4   3 . 4

3 4   3 4 0   0 . 3 4

Example 6

a Work out 0.75×1000.   **b** Find the missing number: $680 \times \ldots = 6.8$

  a Put the digits in columns.

    Multiplying by 1000 moves the digits three places to the left.

    H T U . t h

         0 . 7 5

    7 5 0   Put a 0 in the units column.

    $0.75 \times 1000 = 750$

  b $680 \times \ldots = 6.8$

    H T U . t h   Put the numbers in columns.

    6 8 0

        6 . 8

  The digits have moved two columns to the right. This is multiplication by 0.01.

  The missing number is 0.01. $680 \times 0.01 = 6.8$

Exercise 12D

1 Multiply each number by 10.
   a 32   **b** 6.7   **c** 450   **d** 0.08   **e** 93.75

2 Multiply each number by 100.
   a 60   **b** 0.38   **c** 8.1   **d** 15.7   **e** 1.53

3 Multiply each number by 1000.
   a 0.7   **b** 0.65   **c** 4.3   **d** 1.76   **e** 0.04

4 Copy and complete this table.

	× 10	× 100	× 1000
6.75	67.5		
0.3		30	
4.6			

5 Work these out.
   a 4.8×100 **b** 0.26×1000 **c** 0.27×10 **d** 8.8×1000

6 Multiply each number by 0.1.
   a 6   **b** 140   **c** 73   **d** 19.7   **e** 3.7

7 Multiply each number by 0.01.

    a 7   **b** 29   **c** 200   **d** 4.3   **e** 152

8 Copy and complete this table.

	× 0.1	× 0.01
6	0.6	
38		
290		

9 Work these out.

    a 3.8×0.1 **b** 4.6×100 **c** 15×0.01 **d** 0.08×1000

    e 38×0.01 **f** 0.6×0.1 **g** 2000×0.01 **h** 1.06×100

    i 40×0.001 **j** 1.7×0.01 **k** 0.17×1000 **l** 62.3×100

 10 A sheet of card is 0.1 cm thick. Work out the thickness of 1270 sheets of card.

11 One penny = £0.01.

    Work out the value of 2650 pennies.

12 Copy and complete this table.

	× 10	× 100	× 1000	× 0.1	× 0.01
19		1900			
0.7	7				
6.3					0.063

13 Work out the missing numbers.

    a $3.5 \times \ldots = 350$ **b** $40 \times \ldots = 0.4$ **c** $0.8 \times \ldots = 80$

    d $64.2 \times \ldots = 6.42$ **e** $0.78 \times \ldots = 78$ **f** $3500 \times \ldots = 35$

 14 Which calculation is the odd one out?

    a 360×0.1 **b** 3.6×10 **c** 3600×0.01 **d** 0.36×1000

    Give a reason for your answer.

15 A rectangular strip of metal is 1.7 m long and 0.01 m wide.
Work out the area in square metres (m²).

Challenge: Further multiplication

Look at this multiplication.

$4.3 \times 0.2 = 4.3 \times 0.1 \times 2 = 0.43 \times 2 = 0.86$

Work out these multiplications in the same way.

A 5.4×0.2 **B** 6.1×0.2 **C** 18×0.2 **D** 130×0.2

E 13×0.3 **F** 210×0.4 **G** 44×0.02 **H** 2.4×0.03

I 22×0.4 **J** 34×0.02 **K** 0.6×0.4 **L** 2.4×0.3

12.5 Division with powers of ten

Learning objective

- To mentally divide by a power of ten

Look at these divisions.

$3.4 \div 10 = 0.34$ $3.4 \div 0.1 = 34$ $3.4 \div 0.01 = 340$

H T U . t h   H T U . t h   H T U . t h

    3 . 4   3 . 4   3 . 4
  0 . 3   4   3   4   3   4   0

When you divide by 10 or 100, you move the digits to the right.

When you divide by 0.1 or 0.01, you move the digits to the left.

Compare dividing with the multiplying in section 12.4. You can see that you move the digits in the opposite direction.

Example 7

a Work out $0.65 \div 0.1$

b Find the missing number: $3.1 \div \ldots = 310$

a $0.65 \div 0.1$

H T U . t h

      0 . 6   5
    6 . 5

$0.65 \div 0.1 = 6.5$

Move the digits one place to the left.

b $3.1 \div \ldots = 310$

H T U . t h

        3 . 1
    3   1   0

The missing number is 0.01. $3.1 \div 0.01 = 310$

The digits move two places to the left.

Exercise 12E

1 Divide each number by 10.

  a 5.9   **b** 43   **c** 0.27   **d** 805   **e** 21.7

2 Divide each number by 100.

  a 64   **b** 20.7   **c** 420   **d** 0.5   **e** 1006

3 Work these out.

  a $48 \div 100$ **b** $260 \div 1000$ **c** $0.27 \div 10$ **d** $8.8 \div 10$

  e $423 \div 10$ **f** $4070 \div 100$ **g** $5.2 \div 100$ **h** $12.8 \div 10$

4 Divide each number by 0.1.

  a 9   **b** 23   **c** 0.88   **d** 5.3   **e** 315

5 Divide each number by 0.01.

  a 4   **b** 0.55   **c** 75   **d** 8.3   **e** 12.34

6 Copy and complete this table.

	÷ 10	÷ 0.1	÷ 100	÷ 0.01
310	3.1			
1.7		17		
450				

7 Work these out.

a $4 \div 10$ **b** $4 \div 0.01$ **c** $0.8 \div 0.1$ **d** $0.8 \div 100$

e $4.3 \div 0.1$ **f** $40 \div 1000$ **g** $0.52 \div 0.01$ **h** $0.08 \div 0.001$

 8 A coin is 0.1 cm thick. A pile of these coins is 80 cm high.
How many coins are there in the pile?

 9 Sheets of paper are 0.01 cm thick. How many sheets are in a pile 24 cm high?

 10 a Work these out.

   i $48 \div 10$ **ii** 48×0.1 **iii** $3.9 \div 10$

   iv 3.9×0.1 **v** $620 \div 10$ **vi** 620×0.1

   b What do you notice about your answers to part **a**?

 11 a Work these out.

   i $6 \div 0.1$ **ii** 6×10 **iii** $3.9 \div 0.1$

   iv 3.9×10 **v** $1.58 \div 0.1$ **vi** 1.58×10

   b What do you notice about your answers to part **a**?

12 Find the missing numbers.

a $12 \div \ldots = 120$ **b** $3.6 \div \ldots = 0.36$ **c** $66 \div \ldots = 0.66$ **d** $0.4 \div \ldots = 40$

e $83 \div \ldots = 830$ **f** $1.9 \div \ldots = 0.019$ **g** $4.2 \div \ldots = 0.42$ **h** $27 \div \ldots = 270$

13 Which calculation is the odd one out?

a $620 \div 100$ **b** 62×0.1 **c** 620×0.01 **d** $0.62 \div 0.1$ **e** 0.062×10

Justify your answer.

14 Which calculation is the odd one out?

a $400 \div 100$ **b** 0.4×100 **c** $4 \div 0.1$ **d** 400×0.1 **e** $0.4 \div 0.01$

Justify your answer.

Challenge: Further division

Look at this division.

$6.8 \div 0.2 = 6.8 \div 0.1 \div 2 = 68 \div 2 = 34$

Work out these divisions in the same way.

A $2.4 \div 0.2$ **B** $3 \div 0.2$ **C** $0.84 \div 0.2$ **D** $3.9 \div 0.3$

E $90 \div 0.3$ **F** $0.64 \div 0.4$ **G** $8 \div 0.02$ **H** $0.6 \div 0.03$

I $7 \div 0.5$ **J** $0.32 \div 0.04$ **K** $12 \div 0.6$ **L** $0.09 \div 0.2$

Ready to progress?

I can add and subtract simple fractions.

I can multiply a fraction or a mixed number by an integer.
I can divide an integer by a unit fraction or a unit fraction by an integer.
I can multiply or divide a number by a power of ten.

Review questions

1 Work these out.

a $\frac{3}{8} + \frac{3}{8}$

b $\frac{3}{8} + \frac{1}{2}$

c $\frac{3}{8} + \frac{7}{8}$

d $\frac{3}{8} + \frac{3}{4}$

2 Work these out.

a $\frac{7}{8} - \frac{3}{8}$

b $\frac{7}{8} - \frac{1}{2}$

c $\frac{7}{8} - \frac{1}{4}$

d $\frac{7}{8} - \frac{3}{4}$

3 Work these out.

a $\frac{1}{8} \times 2$

b $\frac{1}{4} \times 7$

c $6 \times \frac{1}{8}$

d $10 \times \frac{1}{5}$

4 Work these out.

a $\frac{1}{5} \div 3$

b $\frac{1}{2} \div 6$

c $\frac{1}{3} \div 4$

d $\frac{1}{4} \div 2$

5 Copy and complete the table.

	× 10	× 100	× 1000
4.23			
0.75			

6 Copy and complete the table.

	÷ 10	÷ 100	÷ 1000
138			
24			

7 Work these out.

a $\frac{2}{3} \times 4$

b $\frac{3}{4} \times 5$

c $6 \times \frac{3}{8}$

d $7 \times \frac{4}{5}$

8 Work these out.

a $5 \times 1\frac{1}{2}$

b $4 \times 3\frac{1}{3}$

c $3 \times 2\frac{3}{4}$

d $4 \times 2\frac{4}{5}$

9 Work these out.

    a $\frac{1}{3} \div 4$ **b** $4 \div \frac{1}{3}$ **c** $3 \div \frac{1}{8}$ **d** $\frac{1}{8} \div 3$

10 Work these out.

    a $8 \div 2$ **b** $2 \div 8$ **c** $\frac{1}{2} \div 8$ **d** $\frac{1}{8} \div 2$ **e** $8 \div \frac{1}{2}$ **f** $2 \div \frac{1}{8}$

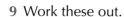
11 A 5p coin is £$\frac{1}{20}$. How many 5p coins make £9?

12 A builder owns 6 hectares of land. He wants to build houses. Each house needs $\frac{1}{25}$ hectare. How many houses can he build?

13 Work these out.

    a 12×0.01 **b** $12 \div 0.1$ **c** 1.2×1000 **d** $0.12 \div 0.1$ **e** 120×0.01

14 Work these out.

    a $36 \div 0.1$ **b** 41.2×100 **c** 900×0.001 **d** $8.3 \div 100$ **e** $17 \div 0.01$

15 a   Work out:

        i $8 \div 0.1$ **ii** $8 \div \frac{1}{10}$

    b   Explain why the answers in part **a** are the same.

16 Work out:

    a $(6 \div \frac{1}{2}) \div 2$ **b** $6 \div (\frac{1}{2} \div 2)$.

17 Work out:

    a $(80 \div 4) \div 0.1$ **b** $80 \div (4 \div 0.1)$.

18 Misha and Lucas are talking about fractions and percentages.

Show how Lucas would explain to Misha that she is not correct.

Problem solving
Making estimates

You may like to work together with a partner on these questions.

Do not use a calculator.

Sometimes it is hard to give an exact answer to a question. However, you can make an estimate. Here are some examples to complete.

1 How much carbon dioxide (CO_2) does a car emit driving from Plymouth to Edinburgh?

The distance from Plymouth to Edinburgh is approximately 800 km.
A typical car emits 100 grams per kilometre of CO_2.

Copy and complete these sentences.

a The amount of CO_2 is …… × …… = …… g.

b 1 kilogram = …… g

c The amount of CO_2 in kilograms is …… ÷ …… = ……

2 A pile of sheets of paper stretches from the floor to the ceiling.
How many sheets are there?

a Estimate the height of the ceiling in centimetres.

b A sheet of paper is about 0.01 cm thick.

Copy and complete this sentence.

The number of sheets is approximately …… ÷ …… = ……

3 How many people could stand on a football pitch?

Pitches vary in size but a typical pitch is about 100 m long and 50 m wide.

a Copy and complete this sentence.

The area of the pitch is …… × …… = …… m^2.

b How many people can fit comfortably in 1 m^2?

c The number of people is …… × …… = ……

 Measure out a square of side 1 metre and estimate how many people could stand inside it.

 Use your answers to a and b.

4 If you left a tap running in your classroom, how long would it take to fill with water?

Assume the tap flows at 10 litres per minute.   1000 litres = 1 m^3

  a Copy and complete this sentence.

    The time to fill 1 m^3 is …… ÷ ……. = …… minutes.

  b Estimate the length, width and height of your classroom in metres, to the nearest metre.

  c Estimate the volume of your classroom in m^3.

    Volume = length × width × height

  d Copy and complete this sentence.

    The time to fill the classroom is …… × …… = …… minutes.

  e Would the room fill up if the tap was running for a whole weekend?

5 If a lift was installed to take you from the Earth to the Moon, how long would it take to get there?

The distance from the Earth to the Moon is about 380 000 km.
Assume the speed of the lift is 100 km per hour.

Time = distance ÷ speed

  a Copy and complete this sentence

    The time to reach the moon is …… ÷ …… = …… hours.

  b A week is 168 hours. Roughly how many weeks will the lift take?

6 How many hairs does the average student have on his or her head?

  a Assume the hairs grow 1 mm apart.
    How many are there in a square of side 1 cm?

    Hint 100 mm^2 = 1 cm^2

  b Estimate the area of a student's scalp in cm^2.

    Hint Imagine the hairs growing in rows.
    How many in a row? How many rows?

  c Copy and complete this sentence.

    The number of hairs is approximately …… × …… = ……

    Hint Imagine it flattened out into a rectangle.

13

Proportion

This chapter is going to show you:

- how to solve problems involving direct proportion
- graphical and algebraic representations of direct proportion
- how to solve problems involving inverse proportion
- graphical and algebraic representations of inverse proportion.

You should already know:

- how to write a simple formula algebraically
- what a ratio is
- how to draw graphs.

About this chapter

The proportions of the parts of the human body change with age.

For an average baby, the length of the head is about a quarter of the length of the whole body.

For an average adult, the length of the head is about an eighth of the length of the whole body.

Ratios of different parts of the body, such as the length of legs compared to the whole body, vary from person to person. Scientists have carried out research to try to find the body ratios that people find most attractive. They have found that the 'ideal' body shape varies from one country to another.

Leonardo da Vinci thought that the ideal proportions should fit in a circle. He made a drawing called *Vitruvian Man* to show this.

13.1 Direct proportion

Learning objectives

- To understand the meaning of direct proportion
- To find missing values in problems involving proportion

Key words

direct proportion

proportional

variable

When you buy petrol you pay a fixed price for each litre. If you buy twice or three times as much, you pay twice or three times as much. This is an example of **direct proportion**.

Two **variables** (such as the number of litres and the price) are in direct proportion if when you multiply one by a number (such as 2 or 3), you multiply the other by the same number.

The same is true of division. When you divide one by a number you divide the other by the same number.

These are some other pairs of variables that are in direct proportion.

- The distance travelled by a car moving at 100 km/hour and the time taken
- The volume of water flowing out of a tap and the time in seconds
- The volume of a fizzy drink and the amount of sweetener in it
- The mass of some loose carrots bought in a shop and the cost in pounds
- The length of time a light has been on and the cost of the electricity used
- The length of a journey in miles and the length of the same journey in kilometres

Example 1

At a petrol station 10 litres of petrol cost £13.60.

Work out the cost of:

a 20 litres   **b** 50 litres   **c** 5 litres.

a You know 10 litres cost £13.60.

20 litres is 10 litres × 2.

The cost is £13.60 × 2 = £27.20.

b 50 litres is 10 litres × 5.   5 × 10 = 50

The cost is £13.60 × 5 = £68.00.

c 5 litres is 10 litres ÷ 2.   10 ÷ 2 = 5

The cost is £13.60 ÷ 2 = £6.80.

You can say that the number of litres and the cost in pounds are directly **proportional**.

Sometimes you leave out the word 'directly' and just say they are proportional.

You will learn about another sort of proportion later in this chapter.

Exercise 13A

 1 200 g of carrots cost 60p.

Work out the cost of:

a 400 g **b** 800 g **c** 600 g **d** 1 kg.

2 250 ml of cola contains 27 g of sugar.

Work out the amount of sugar in:

a 500 ml of cola **b** 750 ml of cola **c** 1 litre of cola.

3 1 kg of flour is enough to make 3 small loaves.

How many small loaves can you make with:

a 2 kg of flour **b** 5 kg of flour **c** 8 kg of flour?

 4 A train is travelling at a constant speed.

In 10 minutes it travels 24 km.

a Work out how far it travels in:

i 20 minutes **ii** 30 minutes **iii** 50 minutes **iv** one hour.

b Copy and complete this table.

Time taken (minutes)	5	10	20	30	45
Distance (km)	12				

c Explain how you know that time taken is directly proportional to distance.

 5 5 miles is approximately the same as 8 km.

a How many kilometres are:

i 10 miles **ii** 15 miles **iii** 20 miles **iv** 30 miles?

b Copy and complete this table. Use the answers to part **a**.

Miles	5	10	15	20	30	40
Kilometres	8					

c Explain how you know that miles are directly proportional to kilometres.

 6 The perimeter of a circle is called the circumference.

The circumference of a circle is proportional to the diameter of the circle.

A circle with a diameter of 7 cm has a circumference of 22 cm.

a Work out the circumference of a circle with a diameter of:

i 14 cm **ii** 21 cm **iii** 28 cm.

b Copy and complete this table. Use the answers to part **a** to help you.

Diameter (cm)	7	14	21	28	35	70
Circumference (cm)	22					

c Explain how you know that the diameter is directly proportional to the circumference.

7 £10 is worth €12.

a How many euros is:

    i £20   **ii** £50   **iii** £80   **iv** £5?

> **Hint** € is the symbol for euros.

b Copy and complete this table. Use the answers to part **a** to help you.

Pounds (£)	10	20	50	80	5	100
Euros (€)	12					

c Explain how you know that pounds (£) are directly proportional to euros (€).

8 The mass of a steel cable is directly proportional to its length.

20 metres of a particular cable has a mass of 6 kg.

Work out the mass of:

a 40 metres   **b** 60 metres   **c** 10 metres   **d** 5 metres.

9 The cost of gold is directly proportional to its mass.

On one day, 12 grams of gold costs £240.

a Work out the cost of:

    i 6 g   **ii** 4 g   **iii** 3 g.

b Copy and complete this table.

Mass of gold (g)	12	6	4	3	2	1
Cost	£240					

10 Water is dripping from a tap at a steady rate.

In one hour (60 minutes) there are 300 drips.

a Work out the number of drips in:

    i 2 hours   **ii** 30 minutes.

b Copy and complete this table.

Time (minutes)	15	20	30	60	120
Number of drips				300	

c Describe the connection between time and the number of drips.

11 In a shop, 100 g of sweets cost 80 pence.

a Use this fact to fill in the table.

Mass (g)	25	50	100	200	500	1000
Cost			80p			

b Work out the cost of:

    i 300 g of sweets   **ii** 500 g of sweets   **iii** 25 g of sweets.

c What mass of sweets can you buy for:

    i £1.60   **ii** £8.00   **iii** 40p?

d Describe the connection between the mass and the cost of the sweets.

12 The pressure of a car tyre can be measured in two different units, bar or psi.

2.1 bar is the same as 30 psi.

Copy and complete this table to show conversions between the two units.

bar	2.1			8.4	12.6
psi	30	10	20		

13 £20 is worth 32 US dollars (US$).

Use this fact to fill in the table.

Pounds	£5	£10	£20	£40	£100
US dollars			$32		

 14 Temperature can be measured in degrees C or degrees F.

Here is a table of values.

Degrees C	20	40	60	80	100
Degrees F	68	104	140	176	212

Explain how you can tell that the temperature in degrees C is not proportional to temperature in degrees F.

Challenge: Average height

This table is from a US website.

It shows the average height of a boy at different ages.

Age (years)	2	4	6	8
Average height (cm)	78	94	107	114

A Average height is not proportional to age. Use numbers from the table to show this.

B Suppose average height is proportional to age.

   a Copy and complete this table to show the heights in this case.

Age (years)	2	4	6	8
Average height (inches)	78			

   b Explain why the numbers in this table are not sensible.

13.2 Graphs and direct proportion

Learning objective

- To represent direct proportion graphically and algebraically

Key word

graph

Look at the table that shows the relationship between distances measured in miles and in kilometres.

They are in direct proportion.

Distance (miles)	10	20	30	50	60	70
Distance (kilometres)	16	32	48	80	96	112

You can plot these values on a **graph** and join them with a line.

There are two things you should notice.

- The points are in a straight line.
- The line passes through the origin.

A graph of values of two variables in direct proportion always has those properties.

Look back at the pairs of values in the table at the start of this section.

Check that:

- $10 \times 1.6 = 16$ $20 \times 1.6 = 32$ $30 \times 1.6 = 48$

and so on.

Number of miles $\times$ 1.6 = number of kilometres

When you know that x miles is the same distance as y kilometres, you can write this as a formula:

$y = 1.6x$

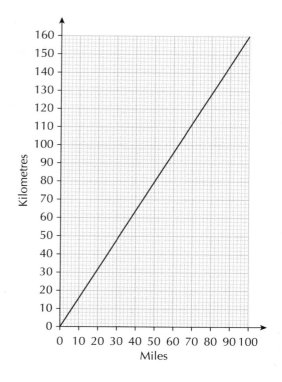

Example 2

Ribbon is sold by the metre. Robena buys 4 metres for 240p.

a Work out the cost of 1 metre.

b Copy and complete this table.

Length (x metres)	1	2	3	4	5
Cost (y pence)				240	

c Draw a graph to show the cost of different lengths of ribbon.

d x metres cost y pence. Copy and complete this formula:

$y = \ldots x$

a 4 metres cost 240p so 1 metre costs $240 \div 4 = 60$p.

b 2 metres cost $2 \times 60 = 120$p, and so on.

Length (x metres)	1	2	3	4	5
Cost (y pence)	60	120	180	240	300

c Put length on the horizontal axis.

Plot the points on graph paper and join them up.

The points are in a straight line. The line goes to the origin.

d 1 metre costs 60p.

x metres cost $x \times 60$ pence.

The formula is $y = 60x$.

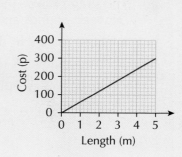

Exercise 13B

1 **a** Each side of an equilateral triangle is 5 cm long.

   Work out the perimeter of the equilateral triangle.

   b Copy and complete this table to show the perimeters of different equilateral triangles.

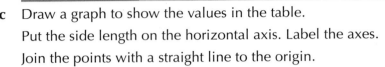

Side length (x cm)	5	10	15	20	25
Perimeter (y cm)					

   c Draw a graph to show the values in the table.

   Put the side length on the horizontal axis. Label the axes.

   Join the points with a straight line to the origin.

   d When the side is x cm the perimeter is y cm.

   Copy and complete this formula: $y = \ldots x$.

2 **a** Each side of a regular pentagon is 10 cm long.

   Work out the perimeter of the pentagon.

   b Copy and complete this table to show the perimeters of different regular pentagons.

Side length (x cm)	10	20	30	40	50
Perimeter (y cm)					

   c Draw a graph to show the values in the table. Put the side length on the horizontal axis. Label the axes.

   d When the side of the pentagon is x cm the perimeter is y cm.

   Copy and complete this formula: $y = \ldots x$.

3 The cost of 1 kg of potatoes is 80p.

   a Copy and complete this table.

Mass (x kg)	1	2	3	4	5
Cost (y pence)	80				

   b Draw a graph to illustrate your figures.

   c Copy and complete this formula for the cost (y pence) of x kg:

   $y = \ldots x$.

4 This graph shows the exchange rate between pounds and Hong Kong dollars.

   a How many HK dollars is £100?

   b How many HK dollars is £50?

   c Copy and complete this table.

Pounds	£50	£100	£150	£200	£250
HK dollars					

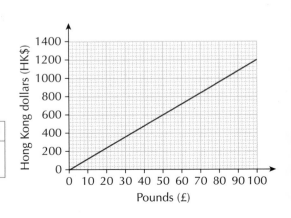

   d What two features of the graph show that the values are proportional?

5 A car is travelling at 70 km/h. The petrol used and distance travelled are shown on this graph.

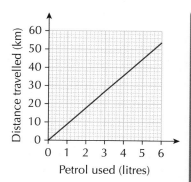

a How far does the car travel on 5 litres of petrol?

b How far does the car travel on 1 litre of petrol?

c Copy and complete this table.

Petrol (x litres)	1	2	3	4	5
Distance (y km)					

d A formula for the distance travelled (y km) on x litres of petrol is $y = \ldots x$.

What is the missing number?

6 The cost of 10 litres of petrol is £13.

a Work out the cost of 20 litres.

b Copy and complete this table.

Petrol (x litres)	10	20	25	30	40
Cost (£y)	13				

c Draw a graph to show the figures in the table. Put litres on the horizontal axis.

d A formula for the cost (£y) of x litres is $y = 1.3x$.

Show that this gives the correct cost of 20 litres.

7 The speed of a car can be measured in metres per second (m/s) or in kilometres per hour (km/h).

5 m/s is the same as 18 km/h.

a Copy and complete this table.

Metres per second (x)	5	10	15	20	25
Kilometres per hour (y)	18			72	

b Use your table to draw a graph.

A formula to convert x m/s to y km/h is $y = 3.6x$.

c Show that the formula give the correct answer when:

i $x = 5$ **ii** $x = 20$.

d Use the formula to convert 40 m/s to km/h.

8 Inches are an old unit for measuring length. 1 inch = 2.5 cm.

a Copy and complete this table.

Inches (x)	1	2	3	4	5	6
Centimetres (y)	2.5					

b Use your table to draw a conversion graph for inches and centimetres.

c A formula to convert x inches to y centimetres is $y = \ldots x$.

Work out the missing number.

d Use your formula to convert 10 inches to centimetres.

9 In the US, people measure their mass in pounds. 10 kg = 22 pounds.

   a Copy and complete this table.

Kilograms (x)	10	20	30	40	50	60
Pounds (y)	22					

   b Use your table to draw a conversion graph for kilograms and pounds.

   c Show that 1 kg = 2.2 pounds.

   d Complete this formula to convert x kilograms to y pounds.

$$y = \ldots x$$

   e Use your formula to convert 55 kg into pounds. Use your graph to check your answer.

10 The angles of this triangle are 30°, 60° and 90°.

The lengths of AB and AC are proportional.

Here is an incomplete table showing possible values of x and y.

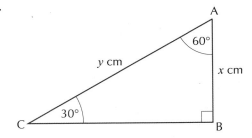

AB (x cm)	2	4	6	8	10
AC (y cm)	4				20

   a Copy and complete the table.

   b Draw a graph to show your values.

   c **i** What do you multiply x by to find y?

      ii Find the missing number in this formula: $y = \ldots x$.

   d Use your formula to find y when:

     i $x = 9$ **ii** $x = 6.5$ **iii** $x = 8.2$.

   e Draw your own triangle with angles of 30°, 60° and 90°, like the one in the diagram. Measure x and y and check that they agree with your formula.

Challenge: Pounds and dollars

This graph shows the exchange rate between pounds (£) and the US dollars ($).

A Use the graph to convert $80 into pounds.

B A formula to convert x into £y is $y = 0.625x$.

   Use this to check your answer to part **A**.

C Use the graph to convert £25 into dollars.

D A formula to covert from £y into x is $x = 1.6y$.

   Use this to check your answer to part **C**.

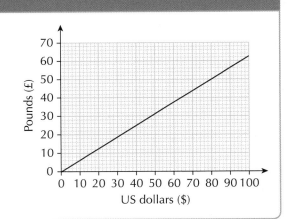

13.3 Inverse proportion

Learning objectives

* To understand what is meant by inverse proportion
* To solve problems using inverse proportion

A car journey is 120 km.

The time it takes depends on the speed of the car.

When the car travels at 60 km/h the journey will take $120 \div 60 = 2$ hours.

When the car travels at 40 km/h the journey will take $120 \div 40 = 3$ hours.

The faster a car travels, the shorter the time taken will be.

The slower a car travels, the longer the time taken will be.

You say that the speed and the time taken are in **inverse proportion**.

The table shows the times at different speeds.

Speed (x km/h)	20	30	40	50	60
Time (y hours)	6	4	3	2.4	2

When you multiply the speed by a number, then you divide the time by the same number.

* speed $= 20 \times 2 = 40$ and time $= 6 \div 2 = 3$
* speed $= 40 \times 1.5 = 60$ and time $= 3 \div 1.5 = 2$

Two variables are in inverse proportion if, when you multiply one by a number you divide the other by the same number.

In this example, when you multiply the speed by the time the answer is always 120 because speed × time = distance.

For example, $20 \times 6 = 120$ or $50 \times 2.4 = 120$.

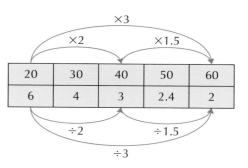

When the speed is x km/h and the time is y hours, you can write this as a formula:

$xy = 120$

When x and y are in inverse proportion you can always write a formula:

$xy = k$

where k is a number.

This graph shows the numbers in the table at the start of this section.

The points are not in a straight line. A smooth curve has been drawn through them.

You can read information from the graph in the same way as you can from a straight-line graph. For example, the curve passes through (10, 12). This tells you that, at a speed of 10 km/h, it would take 12 hours to make the journey.

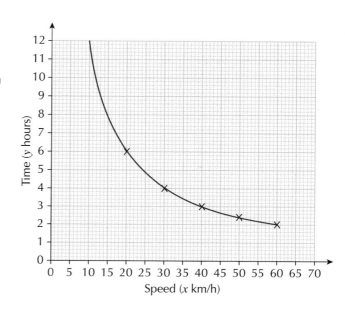

Graphs of inverse proportion are always this shape.

The variables do not have to be x and y. You can use any letters.

Example 3

The area of a rectangular field is 1200 m².

The sides of the field are length a metres and length b metres.

Here is a table showing possible sides of the field.

Length a (metres)	40	50	
Length b (metres)			60

a Find the missing values in the table.

b Work out a formula connecting a and b.

  a The area of the field is the lengths of the two sides multiplied together.

    When $a = 40$: $40 \times b = 1200 \rightarrow b = 1200 \div 40 = 30$

    When $a = 50$: $50 \times b = 1200 \rightarrow b = 1200 \div 50 = 24$

    When $b = 60$: $60 \times a = 1200 \rightarrow a = 1200 \div 60 = 20$

  b The formula is $ab = 1200$.

    In this case the variables, a and b, are inversely proportional.

Example 4

20 horses can survive on a lorry load of hay for 21 days.

a How many days will 1 horse survive for on a lorry load of hay?

b The number of horses and the number of days are inversely proportional.

Copy and complete the table showing how many days different numbers of horses can survive on a lorry load of hay.

Days, d	6	15	21	35	60
Horses, h	70			12	

c Work out a formula connecting h and d.

d How many days can 14 horses survive for on a lorry load of hay?

  a One horse can be fed for $20 \times 21 = 420$ days.

  b When horses and days are inversely proportional, horses × days always equals 420.

    The completed table is:

Days, d	6	15	21	35	60
Horses, h	70	$420 \div 15 = 28$	$420 \div 21 = 20$	12	$420 \div 60 = 7$

  c The formula connecting h and d is $hd = 420$.

  d 14 horses can survive for $420 \div 14 = 30$ days.

Exercise 13C

1 James cycles 4 miles to work in 10 minutes.
  a At this same speed, how far would he travel in 60 minutes?
  b What speed is James cycling at?
  c What is James's speed when he takes 12 minutes to cycle to work?
  d What is James's speed when he takes 6 minutes to cycle to work?

2 Joe eats a bowl of cereal every day. A packet of cereal will last Joe 20 days.
  a How long would it last when Joe and Jess have a bowl of cereal every day?
  b How long would it last when Joe, Jess, Joy and James all have a bowl of cereal every day?

3 A train travels 600 km from Edinburgh to London.
  Use average speed = distance ÷ time to answer these questions.
  a What is the average speed of the train when it takes 6 hours.
  b How long will the train take when it travels at 200 km/h?
  c Copy and complete this table.

Speed, x (km/h)	100	150	120	200	300
Time, y (hours)		4			2

  d Show that speed and time are inversely proportional.
  e Write down a formula connecting x and y from your table.
  f Draw a pair of axes like this.
  g Plot the points from the table in part **c** and join them with a smooth curve.

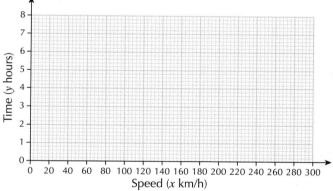

4 A school has £500 to spend on books.
  a How many £10 books can the school buy?
  b How many £50 books can the school buy?
  c Copy and complete this table.

Cost of a book (£x)	2	5	10	20	25	50
Number bought (y)	250	100				

  d Show that x and y are inversely proportional.
  e Write down a formula connecting x and y.
  f Draw a pair of axes like this.
  g Draw a graph to show the information in the table.

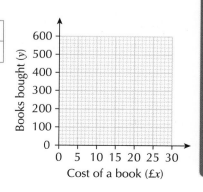

5 The graph shows the time taken by a plane to travel between two airports at different speeds.

   a Copy and complete the table from the information in the graph.

Speed, x (km/h)	400		800	
Journey time, y (hours)	10	8		4

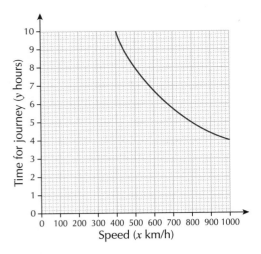

   b Explain how you can tell that the speed of the plane and journey time are inversely proportional.

   c Write down the formula connecting x and y in the table.

6 Children are measuring the length of a pace and how many paces they take to walk 1000 cm.

   a When the length of a pace is 100 cm, how many paces are needed to walk 1000 cm?

   b Find the missing values in this table.

Length of pace, p (cm)	40	50	80	100
Number of paces, n				

   c Show that p and n are inversely proportional.

   d Write down a formula connecting p and n.

7 One large box of cat food is enough to last one cat 90 days.

   a How many days would the box last two cats?

   b How many days would the box last three cats?

   c Is the number of cats and the number of days inversely proportional?

     Explain how you know.

8 A company uses 40 builders to build a block of houses in 36 days.

The number of builders and the numbers of days to build the block is inversely proportional.

   a How many days would it take 20 builders to build the block of houses?

   b How many builders are needed to build the block of houses in 60 days?

   c Copy and complete this table of values.

Number of builders, b	10	20	30	40	60	80
Number of days, d						

   d Draw a graph to show how the number of days to build varies with the number of builders.

   e Write down a formula connecting b and d.

Activity: Different rectangles, same area

A Draw two different rectangles with an area of 36 cm².

B Copy and complete this table showing possible lengths for the height and base of a rectangle with an area of 36 cm².

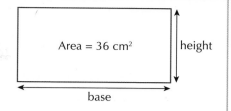

Base, x (cm)	3	4	6		12	
Height, y (cm)	12			4		

C Draw a graph to show your values.

D Write down a formula to show the connection between x and y.

E Use your graph to find the height of a rectangle with an area of 36 cm² and a base of 10 cm.

13.4 The difference between direct proportion and inverse proportion

Learning objectives

- To recognise the difference between direct and inverse proportion in problems
- To work out missing values

You have learnt about direct proportion and inverse proportion. Here is a summary.

Direct proportion

x	1	2	4	10	20
y	10	20	40	100	200

x and y are directly proportional.

When you multiply (or divide) a value of x by a number, you multiply (or divide) y by the same number.

You can always write $y = mx$ where m is a number.

In this example, $y = 10x$.

Inverse proportion

x	1	2	4	10	20
y	100	50	25	10	5

x and y are inversely proportional.

When you multiply a value of x by a number, you divide y by the same number.

You can always write $xy = k$ where k is a number.

In this example $xy = 100$.

Check that each pair of numbers does multiply to make 100.

Example 5

Five bricks have a mass of 45 kg.

a Find the mass of eight bricks.

b How many bricks make a mass of 144 kg?

The higher the number of bricks, the higher the mass. So this is a direct proportion problem.

a Five bricks have a mass of 45 kg.

So one brick will have a mass of $45 \div 5 = 9$ kg.

Eight bricks will have a mass of 8×9 kg $= 72$ kg.

b One brick has a mass of 9 kg.

So a mass of 144 kg will be from $144 \div 9 = 16$ bricks.

Example 6

It takes eight people 12 days to complete a tapestry.

a How long would it have taken six people to complete the tapestry?

b How many people would have been needed to have completed the tapestry in 4 days?

The higher the number of people, the fewer the number of days needed. So this is an inverse proportion problem.

a Eight people complete the tapestry in 12 days.

So one person will complete the tapestry in $12 \times 8 = 96$ days.

Six people will take $96 \div 6 = 16$ days.

b One person takes 96 days.

So to complete the tapestry in 4 days, $96 \div 4 = 24$ people are needed.

Exercise 13D

1 Kathy takes 3 hours to drive to her mother's home when she drives at an average speed of 60 km/h.

a How far does Kathy have to drive to her mother's home?

b How long would it take Kathy if she drove at an average speed of 40 km/h?

c How fast would she need to drive to get there in 2 hours?

2 Eight lumberjacks can fell 24 trees in an afternoon.

a How many trees could 10 lumberjacks fell in an afternoon?

b How many lumberjacks are needed to fell 42 trees in an afternoon?

3 A plane flying at 600 km/h takes 12 hours to complete a journey.

a How far is the journey?

b Another plane flies the same journey at 400 km/h. How long does the journey take?

c Another plane did the same journey in only 10 hours. How fast did it fly?

4 A van uses 70 litres of diesel on a 315 km trip.

   a How much diesel would the van need for a journey of 180 km?

   b How far would the van get if it only had 20 litres of diesel?

5 A group of 15 explorers to the South Pole had just enough food to last them for 40 more days.

   a How long would the food last if all of a sudden three of the explorers disappeared mysteriously?

   b How long would the food last if suddenly, a group of nine other explorers turned up having run out of food?

6 4 tonnes of coal keep a fire burning for 25 weeks.

   a How many tonnes would be needed to keep the fire going for 10 weeks?

   b How many weeks would 5 tonnes of coal keep the fire burning for?

(PS) 7 At an animal reserve, a crate of meat will last six lions 8 weeks.

If the reserve decided to make a crate of meat last for 12 weeks, how many lions will they have to lose?

8 A decorative lamp, filled with 50 ml of oil will keep burning for 6 hours.

   a How long would 30 ml of oil keep the lamp burning for?

   b How much oil is needed to keep the lamp burning for 9 hours?

Reasoning: Looking for proportion

The perimeter of a rectangle is 16 cm.

A When one side of the rectangle is 3 cm, show that the other side is 5 cm.

B Work out three other pairs of values for the two sides of the rectangle.

C Draw a graph to show the pairs of value and draw a line through them.

D Are the lengths of the two sides in direct proportion? Justify your answer.

E Are the lengths of the two sides in inverse proportion? Justify your answer.

Ready to progress?

I can decide when two variables are in direct proportion or inverse proportion.
I can find a formula for two variables in direct proportion.
I can draw a graph for two variables in direct proportion.
I can find a formula connecting two variables in inverse proportion.
I can draw a graph for two variables in inverse proportion.

Review questions

1   A wire is 12 cm long and has a mass of 3 g.

Work out the mass of a length of wire that is:

   a   24 cm long   b   36 cm long   c   60 cm long.

FS 2   Tom can earn £1500 a week when he works six days a week.

   a   How much will Tom earn when he only works three days a week?

   b   Tom earns £1000 one week. How many days did Tom work?

   c   What is the most Tom can earn in a week?

MR 3   A coach on a motorway is travelling at a constant speed.

In 10 minutes it travels 16 km.

   a   Work out how far it travels in:

      i   20 minutes   ii   30 minutes   iii   50 minutes   iv   one hour.

   b   Copy and complete this table.

Time taken (minutes)	5	10	20	30	45
Distance (km)		16			

   c   Explain what type of proportion there is between time taken and distance.

FS 4   In a market, 100 g of raspberries cost £1.50.

   a   Use this fact to copy and complete the table.

Mass	10 g	50 g	100 g	200 g	500 g	1000 g
Cost			£1.50			

   b   Work out the cost of:

      i   200 g of raspberries   ii   250 g of raspberries   iii   125 g of raspberries.

   c   What mass of raspberries can you buy for:

      i   £1.20   ii   £2.40   iii   £6.60.

   d   Explain what type of proportion there is between mass and cost of raspberries.

5 a Each side of a regular heptagon is 5 cm long.

Work out the perimeter of the heptagon.

b Copy and complete this table to show the perimeters of different regular heptagons.

5 cm

Length of side, x (cm)	5	10	15	20
Perimeter, y (cm)				

c Draw a graph to show the values in the table. Put length of side on the horizontal axis. Label the axes.

d The side of the pentagon is x cm and the perimeter is y cm.

Copy and complete this formula:

$y = \ldots x.$

 6 Racing cars were tested to find the petrol consumption when travelling at different speeds. This graph was drawn from the results.

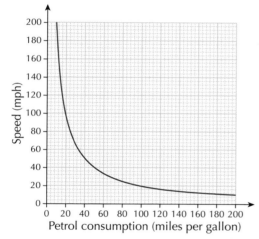

a Copy and complete the table from information in the graph.

Petrol consumption (mpg)	10		40	50	100
Speed (mph)	200	100			

b Explain how you can tell that the speed of the racing car and the petrol consumption are inversely proportional.

PS 7 It takes eight people three days to harvest all the potatoes in a farmer's field.

How long would it have taken six people?

PS 8 Six men spend 12 days building a house.

a How long would it have taken four men to build the same house?

b How many men would have been needed to have built the house in nine days?

Challenge
Coach trip

A youth leader is planning a coach trip for some youth club members.

1 Cost of the coach

The total cost of the coach is £600.

This must be shared equally by the people on the coach.

a Copy and complete this table.

Number of passengers (x)	50	40	32	25	20
Cost for each person (£y)	12				30

b Show that the number of passengers and the cost per person are in inverse proportion.

c Draw a graph to show how the cost per person varies with the number of people.

2 Time and distance

On the motorway the coach travels at a constant speed of 80 km/h.

a How far does the coach travel in half an hour (30 minutes) on the motorway?

b How far does the coach travel in 10 minutes on the motorway?

c Copy and complete this table.

Time (minutes)	10	20	30	40	50	60
Distance (km)						

d Is the distance proportional to the time? Explain your answer.

e Draw a graph to show the data in the table.

3 Speed and time

The time taken to complete the whole journey is inversely proportional to the average speed.

a Copy and complete this table.

Average speed (x km/h)	50	60	80	100
Time (y hours)			3	

b Write down a formula connecting x and y.

c Draw a graph to show how the time varies with the average speed.

4 Fuel consumption

The amount of fuel the coach uses varies with the distance.

Look at the data in this table. It shows the amounts of fuel used when the coach is travelling on the motorway at 80 km/h.

Fuel used (f litres)	25	50	80
Distance travelled (d km)	75	150	240

a Is the relationship between fuel used and distance travelled:

   i direct proportion
   ii inverse proportion
   iii neither of these?

b Use your answer to part **a** to write a formula for d in terms of f.

The fuel consumption of the coach is measured in litres per kilometre (litres/km). The fuel consumption varies with speed.

Look at the data in this table.

Speed (km/h)	40	60	80
Fuel consumption (litres/km)	3	2.5	2

c Is the relationship between speed and fuel consumption:

   i direct proportion
   ii inverse proportion
   iii neither of these?

Give a reason for your answer.

14

Circles

This chapter is going to show you:

- the names of the parts of a circle
- how to work out the circumference of a circle by measuring
- how to use a formula to work out the approximate length of the circumference.

You should already know:

- the words 'radius' and 'diameter'
- how to use compasses to draw a circle.

About this chapter

Circles are all around us.

The circle is probably the most important shape in the universe.

Look at buildings. They are made up of lots of shapes, including circles. Modern buildings such as stadiums and shopping complexes are built so that they look attractive. They contain a vast mixture of symmetrical shapes, including circles.

14.1 The circle and its parts

Learning objective

- To know the definition of a circle and the names of its parts

A circle is a set of points that are all the same distance from a fixed point, called the **centre**.

You must learn all of these words for the different parts of a circle.

- The centre of the circle is usually called O.

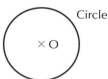

- **Arc**: a part of the circumference of the circle.

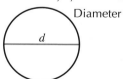

- **Diameter**: the distance across a circle, through its centre. The diameter, d, of a circle is twice its radius, r, so $d = 2r$.

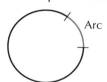

- **Sector**: a portion of a circle enclosed by two radii and an arc.

- **Circumference**: the length round the outside of the circle. It is a special name for the perimeter of a circle.

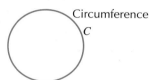

- **Radius**: the distance from the centre of a circle to its circumference. The plural of radius is *radii*.

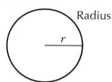

- **Chord**: a straight line that joins two points on the circumference of a circle. If the chord passes through the centre, it is also a diameter.

- **Semicircle**: one half of a circle; either of the parts cut off by a diameter.

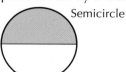

Example 1

a i Measure the radius of this circle, giving your answer in centimetres.

ii Write down the diameter of the circle.

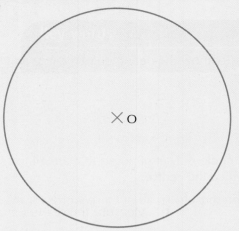

b i Measure the diameter of this circle, giving your answer in millimetres.

ii Write down the radius of the circle.

   a i radius = 3 cm

     ii diameter = radius × 2 = 6 cm

   b i diameter = 40 mm

     ii radius = diameter ÷ 2 = 20 mm

Exercise 14A

1 Write down the name of each part of the circle.
Use words from this list.

> arc   centre   chord   circumference
> diameter   radius   sector   semicircle

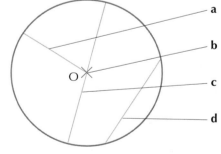

2 Write down the name of each part of the circle.
Use words from this list.

> arc   centre   chord   circumference
> diameter   radius   sector   semicircle

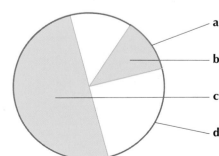

3 **i** Measure the radius of each circle, giving your answer in centimetres.

  ii Write down the diameter of each circle.

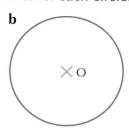
a

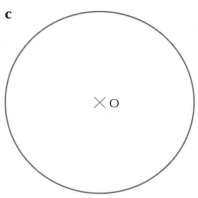
b   **c**

4 **i** Measure the diameter of each circle, giving your answer in millimetres.

  ii Write down the radius of each circle.

a   **b**   **c**

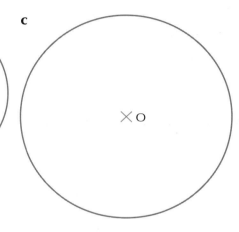

5 Draw circles with these measurements.

  a radius = 2 cm   **b** radius = 3.5 cm   **c** diameter = 8 cm   **d** diameter = 10 cm

(MR) **6** Draw a circle with a radius of 4 cm. On your circle draw four chords with lengths 4 cm, 5 cm, 6 cm and 7 cm. Your chords must not cross or touch each other.

(MR) **7** Draw each shape accurately. Use a ruler, compasses and a protractor.

a   **b**

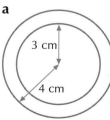
3 cm

4 cm

Concentric circles

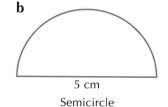
5 cm

Semicircle

c   **d**

3 cm

3 cm

Quadrant of
a circle

4 cm 60° 4 cm

Sector of
a circle

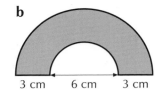

8 Draw each shape accurately.

a

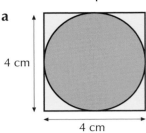

4 cm

4 cm

b

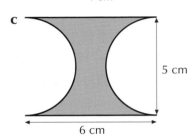

3 cm   6 cm   3 cm

c

5 cm

6 cm

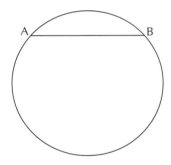

Activity: How to find the centre of a circle

A Draw a circle around a circular object so that you do not know where the centre is.

B Draw a chord AB on the circle.

C Mark a point X halfway between A and B.

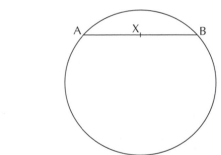

D Use a protractor to draw a line that is perpendicular to the chord AB and that passes through X.

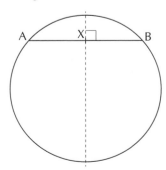

E Now repeat these steps for a different chord CD with Y halfway between C and D.

Where the two perpendiculars cross is the centre of the circle, O.

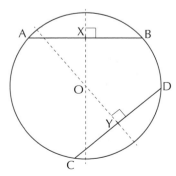

14.2 Circumference of a circle

Learning objective

- To work out the relationship between the circumference and diameter of a circle

How can you measure the circumference of a circle exactly?

Is there a relationship between the length of the diameter and the circumference?

Exercise 14B will show you.

Exercise 14B

You will need compasses, a 30 cm ruler and a piece of fine, high-quality string at least 40 cm long.

Copy this table and then draw circles with the given radii.

Radius r (cm)	Diameter d (cm)	Circumference C (cm)	$C \div d$
1			
2			
3			
4			
5			
6			

Measure the circumference of each circle by placing the string round the circumference, as shown. Make a pencil mark on the string where it meets its starting point.

Use the ruler to measure this length. Complete the table.

Calculate the value to write in the last column. Give each answer correct to one decimal place.

What do you notice about the numbers in the last column?

How is the circumference related to the diameter?

Write down in your book what you have found out.

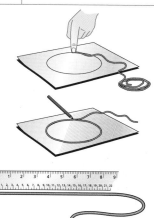

Activity: Making nets for cones

A Draw a circle with radius 4 cm on paper and cut it out.

B Draw a narrow sector on the circle and cut it out.

C Make a cone with the remaining larger sector.

D Now repeat the activity for two more circles with the same radius but cut out larger sectors on each one.

E What happens as you increase the size of the sector you remove?

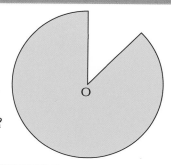

14.3 A formula to work out the approximate circumference of a circle

Learning objective

- To use a formula to work out the approximate circumference of a circle

In Exercise 14B, you should have found that the circumference, C, of a circle with diameter d, is given approximately by the formula $C = 3d$. The ancient Chinese were the first to use this formula.

You can use this formula to work out the approximate length of the circumference of any circle.

Example 2

Work out the approximate circumference of this circle.

The diameter $d = 6$ cm, which gives:

$C = 3d = 3 \times 6 = 18$ cm

Example 3

Work out the approximate circumference of this circle.

Here you are given the radius and the units are metres.

$r = 3.4$ m so $d = 3.4 \times 2 = 6.8$ m

This gives:

$C = 3d = 3 \times 6.8 = 20.4$ m

Exercise 14C

1 Work out the approximate circumference of each circle.

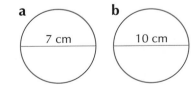

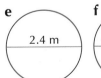

a   7 cm   b   10 cm   c   4.5 cm   d   8.5 cm   e   2.4 m   f   5.6 m

2 Work out the approximate circumference of each circle.

a   4 cm   b   4.5 cm   c   7 cm   d   1.4 m   e   3.6 m   f   5.5 m

3 The diameter of a £2 coin is 28 mm.

Work out the approximate circumference of the coin.

MR **4** The 'High Roller' in Las Vegas has a diameter of 160 m.

    a How far would you travel, approximately, in one complete revolution of the wheel?

    b The wheel makes three complete revolutions.

Would you travel more than one kilometre?

Explain your answer.

 Hint   1 km = 1000 m

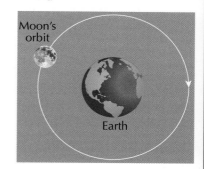

5 The moon's orbit around the Earth can be taken to be a circle with a radius of approximately 384 000 kilometres.

Work out the approximate distance that the moon travels in one orbit of the Earth. Give your answer correct to the nearest million kilometres.

Moon's orbit

Earth

6 A circle has a circumference of 24 cm.

Work out the approximate diameter of the circle.

PS **7** A circle has a circumference of 4.8 m.

Work out the approximate radius of the circle.

Give your answer in centimetres.

Problem solving: Working out areas

A This tin has a diameter of 7 cm.

The height of the label around the tin is 10 cm.

Work out the approximate length of the label if it does not overlap.

Then work out the area of the rectangular label.

 **Hint**   Work out the circumference of the circle first.

SPAGHETTI HOOPS

B This tin has a radius of 4.5 cm.

The height of the label around the tin is 4 cm.

Work out the approximate area of the label when it has an overlap of 1 cm.

Red Salmon

Ready to progress?

I know the names of the parts of a circle.
I can measure the radius and diameter of a circle.

I can use a formula to work out the approximate circumference of a circle.

Review questions

1 Draw a circle with a radius of 5 cm and centre at O.

  On the circle draw and label:

  a a radius   b a diameter   c a chord   d a sector.

2 i Measure the diameter of each circle, giving your answer in centimetres.

  ii Write down the radius of each circle.

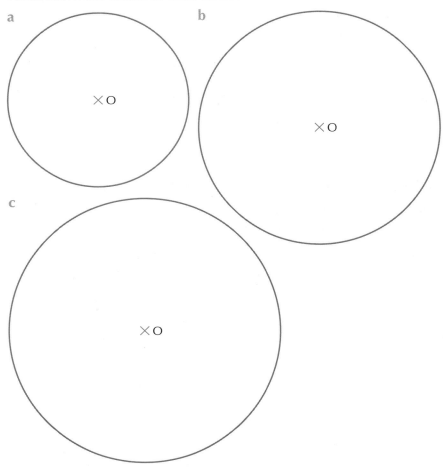

3   Draw circles with these measurements:

   a   radius = 2.5 cm   b   radius = 3.5 cm   c   diameter = 10 cm.

4   Work out the approximate circumference of each circle.

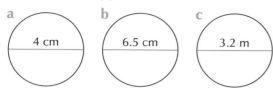

a   4 cm   b   6.5 cm   c   3.2 m

5   Work out the approximate circumference of each circle.

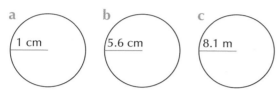

a   1 cm   b   5.6 cm   c   8.1 m

6   The circumference of this circular running track is 200 m.

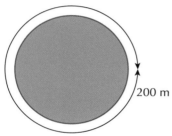

200 m

   Work out the approximate diameter of the track.

   Give your answer to the nearest metre.

7   The wheels of a bicycle have a diameter of 65 cm.

   a   Work out the approximate circumference of each wheel.

       Give your answer in metres.

   b   Lyndon cycles a distance of 5 km.

       Work out the approximate number of times the wheels turn.
       Give your answer to the nearest 10.

Activity
Constructions

The following are three important geometric constructions.
They are useful because they give exact measurements and are
therefore used by architects and in design and technology.

You will need a sharp pencil, a ruler, compasses and a protractor.

Leave all your construction lines on the diagrams.

Example 1 To construct the mid-point of AB and the perpendicular to the line AB.

- Draw a line AB of any length.

- Set your compasses to any radius greater than half the length of AB.

- Draw two arcs with the centre at A,
  one above and one below AB.

- With compasses set at the same radius,
  draw two arcs with the centre at B, to
  intersect the first two arcs at C and D.

- Join C and D to intersect AB at X.

- X is the mid-point of the line AB.

- The line CD is the perpendicular to the line AB.

- The line CD is often referred to as the
  perpendicular bisector to the line AB.

1 Draw a line AB 6 cm in length. Using compasses, construct the perpendicular
   bisector of the line.

2 Draw a line CD 7.5 cm in length. Using compasses, construct the perpendicular
   bisector of the line.

Example 2 To construct an isosceles triangle XYZ with perpendicular height 7 cm.

- Draw a line XY 5 cm in length.

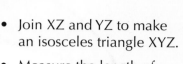

- Construct the perpendicular bisector of XY.

- Make the perpendicular bisector 7 cm long to find the point Z.

- Join XZ and YZ to make an isosceles triangle XYZ.

- Measure the length of XZ and YZ.

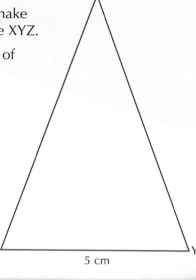

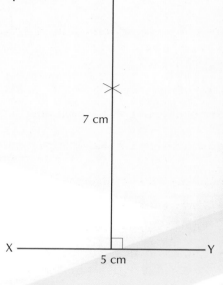

Example 3 To construct the bisector of the angle ABC.

- Draw an angle ABC of any size.

- Set compasses to any radius. With the centre at B, draw an arc to intersect BC at X and AB at Y.

- With compasses set to any radius, draw two arcs with the centres at X and Y, to intersect at Z.

- Join BZ.

- BZ is the bisector of the angle ABC.

- Then angle ABZ = angle CBZ.

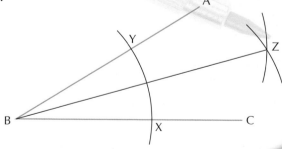

3 Using a protractor draw an angle of 60°.
   Using compasses, construct the angle bisector of this angle.
   Measure the two angles formed to check that they are both 30°.

4 Using a protractor draw an angle of 120°.
   Using compasses, construct the angle bisector of this angle.
   Measure the two angles formed to check that they are both 60°.

15

Equations and formulae

This chapter is going to show you:

- how to solve equations
- how to substitute values into a variety of formulae.

You should already know:

- how to use simple algebra.

About this chapter

Behind most of today's technology lie equations, whether it is a space probe going into orbit or a computerised supermarket cash till. When scientists plan a mission, such as taking a spacecraft to Mars or trying to land on a comet, they have to carry out extremely detailed plans for the mission. They use formulae and equations to simulate the path the spacecraft will take, and try out different paths to see which is the best one. They also use equations to see what effect small changes will have on this path and to try to model things that could go wrong. You'll be pleased to read that these equations are much more complicated than any you will meet in this book.

15.1 Equations

Learning objective

- To solve simple equations

Key words

equation

inverse operation

solve

You have learnt how to **solve equations** by using **inverse operations**.

For example:

- the inverse of subtract 2 is add 2
- the inverse of multiply by 3 is divide by 3.

Andrew thinks of a number.

He adds 7 to the number.

Suppose Andrew's number was n.

When you add 7 then the result is $n + 7$.

Andrew says that the answer is 12.

What number did Andrew start with?

You can see that it is $n = 5$ because $5 + 7 = 12$.

This is called solving an equation.

Example 1

Sophia thinks of a number.

She subtracts 8.

The answer is 12.

a Call Sophia's number N.

Write down an expression to show what Sophia did.

b Find the value of N.

   a When you subtract 8 from N you get $N - 8$.

   b Now solve $N - 8 = 12$.

$$N - 8 + 8 = 12 + 8 \qquad \text{Add 8 to both sides.}$$
$$N = 20$$

Check with $20 - 8 = 12$.

Example 2

Solve these equations.

a $3 + x = 8$ **b** $4x = 24$ **c** $\dfrac{x}{2} = 7$ **d** $x - 3 = 12$ **e** $x + 5 = 2$

a $3 + x = 8$ Subtract 3 from both sides to get x on its own.

$\qquad x = 5$

b $\qquad 4x = 24$ Divide both sides by 4, to get x on its own.

$\quad 4x \div 4 = 24 \div 4$

$\qquad\qquad x = 6$

c $\dfrac{x}{2} = 7$ Multiply both sides by 2 to get x on its own.

$\dfrac{x}{2} \times 2 = 7 \times 2$

$\qquad x = 14$

d $x - 3 = 12$ Add 3 to both sides to get x on its own.

$\quad x - 3 + 3 = 12 + 3$

$\qquad\qquad x = 15$

e $\qquad x + 5 = 2$ Subtract 5 from both sides to get x on its own.

$\quad x + 5 - 5 = 2 - 5$

$\qquad\qquad x = -3$

Example 3

a Write down an expression for the perimeter of this triangle.

b The perimeter of the triangle is 30 cm.

Write down an equation about the perimeter involving x.

c Solve this equation.

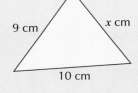

9 cm x cm   10 cm

a Perimeter $= x + 9 + 10$ Collect like terms.

$\qquad\qquad = x + 19$

b As the perimeter is 30 cm, the equation is $x + 19 = 30$.

c $x + 19 = 30$ Subtract 19 from both sides.

$\qquad x = 11$

Example 4

a Write down an expression for the area of this rectangle.

b The area of this rectangle is 35 cm^2.

Write down an equation about the area involving y.

c Solve this equation.

y cm

5 cm

a The area is $5 \times y$ which can be written as $5y$.

b As the area is 35 cm^2, then $5y = 35$.

c $5y = 35$ Divide both sides by 5.

$\quad y = 7$

Exercise 15A

1 Solve each of these equations.

   a $x + 1 = 7$ **b** $x - 2 = 7$ **c** $x + 4 = 8$ **d** $x - 6 = 5$

   e $m + 5 = 11$ **f** $m - 2 = 2$ **g** $k + 3 = 5$ **h** $p - 1 = 10$

   i $k + 5 = 15$ **j** $k + 5 = 9$ **k** $m - 1 = 1$ **l** $x - 4 = 2$

2 Solve each of these equations.

   a $2x = 8$ **b** $3x = 15$ **c** $4x = 20$ **d** $5x = 45$

   e $3m = 12$ **f** $5m = 30$ **g** $6m = 42$ **h** $7m = 28$

   i $3k = 27$ **j** $7k = 35$ **k** $8k = 24$ **l** $2k = 22$

3 Solve each of these equations.

   a $\frac{x}{2} = 8$ **b** $\frac{x}{9} = 2$ **c** $\frac{x}{3} = 2$ **d** $\frac{x}{5} = 3$

   e $\frac{x}{2} = 7$ **f** $\frac{x}{4} = 1$ **g** $\frac{x}{3} = 8$ **h** $\frac{x}{4} = 6$

4 Solve each of these equations.

   a $x + 5 = 4$ **b** $x + 6 = 2$ **c** $x + 4 = 1$

   d $x + 6 = 5$ **e** $m + 5 = 7$ **f** $m + 2 = 1$

   g $k + 8 = 5$ **h** $p + 1 = 9$ **i** $k + 15 = 5$

   j $k + 9 = 4$ **k** $m + 3 = 7$ **l** $x + 7 = 2$

> **Hint**   Many of the equations have negative answers.

5 In these algebraic brick walls, each number is the sum of the numbers in the two bricks below. Find the unknown number x in each case.

a

b

c

d

e

f

g
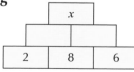

h
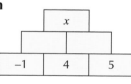

i
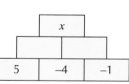

j
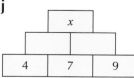

k
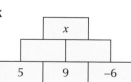

l
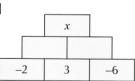

6 In these algebraic brick walls, each number is the sum of the numbers in the two bricks below. Find the unknown number x in each case.

a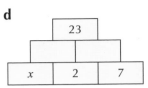

b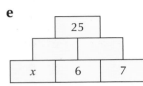

c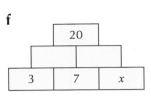

d

e

f

7 **a** Write down an expression for the perimeter of this triangle.
  b The perimeter of the triangle is 20 cm.
        Write down an equation involving t.
  c Solve the equation.

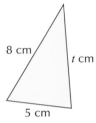

8 **a** Write down an expression for the area of this rectangle.
  b The area of the rectangle is 72 cm^2.
        Write down an equation involving x.
  c Solve the equation.

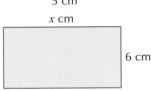

Challenge: Odd one out

Using only whole positive numbers, in how many different ways can you complete each of these? Show all your answers.

A …+…= 9   **B** …×…= 24   **C** …−…= 7

15.2 Equations with brackets

Learning objective

• To solve equations which include brackets

You should know how to multiply out brackets in algebraic expressions.

Eve is thinking of a number.

She adds 3 to it and then doubles the answer.

Call Eve's number x.

She adds 3 to x to get $x + 3$.

She then doubles it to get $2(x + 3)$.

She now says that the final answer is 18.

What number did she start with?

You can write the equation:

$2(x + 3) = 18$

This equation can now be solved as shown in the example.

Example 5

Solve the equation $2(x + 3) = 18$.

$2(x + 3) = 18$	Divide both sides by 2.	$18 \div 2 = 9$
$x + 3 = 9$	Subtract 3 from both sides.	
$x = 6$		

Exercise 15B

1   Solve each equation.

   a $x + 5 = 12$ **b** $x - 3 = 12$ **c** $x + 17 = 7$ **d** $x - 2 = 11$

   e $m - 5 = 1$ **f** $k + 5 = 14$ **g** $n - 6 = 4$ **h** $x - 6 = 0$

2   Solve each equation.

   a $2(x + 1) = 14$ **b** $3(m - 6) = 12$ **c** $3(x - 4) = 15$ **d** $5(t + 2) = 25$

   e $2(x + 8) = 20$ **f** $2(m - 5) = 6$ **g** $4(m - 1) = 12$ **h** $6(x - 3) = 24$

   i $4(m - 2) = 16$ **j** $2(k + 4) = 10$ **k** $6(x - 5) = 18$ **l** $2(t - 3) = 8$

   m $3(x - 7) = 21$ **n** $8(y - 9) = 24$ **o** $3(x + 4) = 27$ **p** $3(m - 1) = 33$

3   Solve each equation.

    Many solutions are negative.

   a $x + 5 = 1$ **b** $m + 6 = 4$ **c** $x + 4 = 3$ **d** $t + 2 = 5$

   e $2(w + 5) = 6$ **f** $2(t + 4) = 4$ **g** $4(n + 7) = 8$ **h** $6(y + 5) = 18$

   i $4(x + 4) = 12$ **j** $2(h + 7) = 12$ **k** $6(g + 5) = 24$ **l** $2(q + 3) = 18$

   m $3(n + 7) = 15$ **n** $8(t + 7) = 32$ **o** $3(y + 5) = 9$ **p** $5(p + 8) = 35$

 4   The perimeter of this rectangle is calculated using the equation:

perimeter $= 2(x + 8)$

8 cm

x cm

   a   **i** Write an equation involving x when the perimeter is 30 cm.

       ii Solve this equation.

   b   **i** Write an equation involving x when the perimeter is 48 cm.

       ii Solve this equation.

   c   The area of the rectangle is 40 cm^2.

       i Write an equation involving x.

       ii Solve this equation.

   d   What is the perimeter of this rectangle when the area is 72 cm^2?

5 You can calculate the area of this trapezium using the formula:

area = 4(10 + x)

a   i Write an equation involving x when the area is 68 cm².

   ii Solve this equation.

b   i Write an equation involving x when the area is 56 cm².

   ii Solve this equation.

c   The area of the trapezium is now 48 cm².

   i Write an equation involving x.

   ii Solve this equation.

d   What is the value of x in this trapezium when the area is 72 cm²?

  6 The sum of the interior angles of a polygon with n sides is given by the rule:

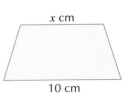

angle sum = 180(n − 2)°

a   i Write an equation involving n when the angle sum of a polygon is 360°.

   ii Solve this equation.

   iii What we call a polygon with an angle sum of 360°?

b   i Write an equation involving n when the angle sum is 720°.

   ii Solve this equation.

   iii What is the name of a polygon with an angle sum of 720°?

c   The angle sum of a polygon is 540°.

   i Write an equation involving n.

   ii Solve this equation.

   iii What is the name of a polygon with an angle sum of 540°?

d   What is the name of a polygon with an angle sum of 1080°?

Challenge: Muddying the waters

The triangular numbers are given by the formula:

$$T = \frac{n(n + 1)}{2}$$

You calculate the first triangular number by substituting in $n = 1$, which gives:

$$T = \frac{1 \times (1 + 1)}{2} = 1$$

To find the second triangular number, substitute $n = 2$, which gives:

$$T = \frac{2 \times (2 + 1)}{2} = 3$$

A   Find the next four triangular numbers.

B   Find the 99th triangular number.

15.3 More complex equations

Learning objective

• To solve equations involving two operations

The equations you started solving involved just one operation such as addition, subtraction, multiplication or division, for example, as $x + 3 = 7$ or $4x = 12$.

Then you solved equations that involved two operations such as $3n + 2 = 14$ where the n is multiplied by 3 then 2 is added to give the sum of 14.

You have also solved equations involving brackets which gave two operations to solve, getting the term inside the brackets on its own by dividing by the number outside the brackets. Then you solved the simple equation you were left with.

You have also solved simple equations with fractions.

In this section, you will solve more complex equations with fractions.

Example 6

Solve the equation $3t + 5 = 17$.

$3t + 5 = 17$ Subtract 5 from both sides to get the term in t on its own. $17 - 5 = 12$.

$3t = 12$ Divide both sides by 3.

$t = 4$

You can check this is correct by substituting $t = 4$ into the equation.

$3 \times 4 + 5 = 12 + 5 = 17$

Example 7

Solve the equation $5m - 2 = 13$.

$5m - 2 = 13$ Add 2 to both sides.

$5m = 15$ Divide both sides by 5.

$m = 3$

You can check this is correct by substituting $m = 3$ into the equation.

$5 \times 3 - 2 = 15 - 2 = 13$

Example 8

Solve the equation $\frac{x}{3} + 4 = 11$.

$\frac{x}{3} + 4 = 11$ First get the term in x on its own. Subtract 4 from both sides.

$\frac{x}{3} = 7$ Multiply both sides by 3. $7 \times 3 = 21$

$x = 21$

You can check this is correct by substituting $x = 21$ into the equation.

$21 \div 3 + 4 = 7 + 4 = 11$

Exercise 15C

1 Solve each equation.

a $2x + 3 = 15$ **b** $4x + 1 = 29$ **c** $3x + 4 = 25$

d $5x + 1 = 11$ **e** $2x + 3 = 13$ **f** $4x + 3 = 23$

2 Solve each equation.

a $2(t + 4) = 16$ **b** $4(y + 3) = 28$ **c** $3(m + 5) = 24$

d $5(n + 3) = 20$ **e** $2(t + 4) = 12$ **f** $4(x + 1) = 32$

3 Solve each equation.

a $2x - 3 = 17$ **b** $4x - 1 = 19$ **c** $5x + 3 = 18$ **d** $2y - 3 = 13$

e $4t + 5 = 17$ **f** $6x - 5 = 13$ **g** $10b + 9 = 29$ **h** $2r - 3 = 9$

i $3x - 11 = 1$ **j** $7p + 5 = 40$ **k** $5x + 7 = 52$ **l** $9x - 8 = 55$

4 Solve each equation.

a $2m + 3 = 1$ **b** $4t + 5 = 1$ **c** $2m + 8 = 4$

d $2k + 7 = 1$ **e** $5x + 9 = 4$ **f** $6y + 5 = 11$

g $3a + 9 = 3$ **h** $2n + 9 = 3$ **i** $3x + 11 = 2$

> **Hint** Many answers are negative.

5 Solve these equations.

a $\dfrac{k}{3} + 3 = 8$ **b** $\dfrac{x}{3} + 3 = 6$ **c** $\dfrac{m}{5} - 3 = 2$ **d** $\dfrac{x}{4} + 4 = 6$

e $\dfrac{x}{4} - 2 = 5$ **f** $\dfrac{m}{5} + 3 = 5$ **g** $\dfrac{k}{6} - 3 = 4$ **h** $\dfrac{m}{4} - 3 = 2$

i $\dfrac{k}{6} + 3 = 7$ **j** $\dfrac{x}{4} - 5 = 3$ **k** $\dfrac{k}{3} - 2 = 4$ **l** $\dfrac{m}{5} + 5 = 11$

6 Solve each equation.

a $2p - 1 = 19$ **b** $4(m + 3) = 16$ **c** $\dfrac{m}{5} + 3 = 8$

d $2t + 8 = 2$ **e** $4(x - 5) = 32$ **f** $\dfrac{x}{2} - 5 = 3$

g $8a + 9 = 33$ **h** $2(k + 5) = 18$ **i** $\dfrac{x}{3} - 7 = 8$

7 For each of these number walls:

i use the number at the top to write an equation involving x

ii solve the equation.

The first one has been done for you.

a

16

| $x + 3$ | $x + 7$ |

b

21

| $x + 2$ | $x + 5$ |

c

17

| x | $x + 3$ |

i	$x + 3 + x + 7 = 16$
	$2x + 10 = 16$
ii	$2x = 6$
	$x = 3$

8 For each number wall:

i use the number at the top to write an equation involving x

ii solve the equation.

a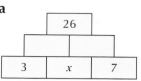

b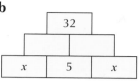

c

(MR) **9** This is a page from Terry's exercise book.

1. $2m + 3 = 11$
    $2m = 14$
    $m = 7$

2. $3(t - 1) = 12$
    $t - 1 = 4$
    $t = 3$

3. $\frac{x}{2} - 3 = 5$
    $\frac{x}{2} = 8$
    $x = 4$

4. $5(x + 3) = 25$
    $x + 3 = 20$
    $x = 17$

He has made a mistake in each question.

Explain what he has done wrong in each one and correct it for him.

Challenge: Think of a number

Pete had a number rule in his head.

He used the rule on any number given to him.

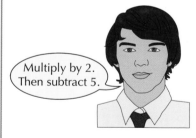

Multiply by 2.
Then subtract 5.

A What number did Pete give in reply to each of the following?

a 7 **b** 3 **c** 10

B What number was given to Pete when he gave these replies?

a 19 **b** 9 **c** 21

15.4 Substituting into formulae

Learning objective

• To substitute values into a variety of formulae

Formulae occur in all sorts of situations. A common example is the conversion between metric and imperial units.

Example 9

The formula for converting kilograms (K) to pounds (P) is:

$P = 2.2K$

Convert 10 kilograms to pounds.

Substituting $K = 10$ into the formula gives:

$P = 2.2 \times 10 = 22$

So, 10 kg is 22 pounds.

Example 10

The formula for the volume, V, of a box with length b, width w and height h, is given by:

$V = bwh$

Calculate the volume of a box whose length is 8 cm, width is 3 cm and height is 2 cm.

Substitute the values given into the formula:

$V = 8 \times 3 \times 2 = 48$

So, the volume of the box is 48 cm^3.

Exercise 15D

1 The area, A, of a rectangle is found using the formula $A = LB$, where L is its length and B is its width.

Find the area, A, when:

a $L = 8$ and $B = 7$ **b** $L = 6$ and $B = 1.5$.

2 The surface area, A, of a cylinder with no ends is given by the formula $A = 6rh$.

Find the surface area, A, when:

a $r = 6$ and $h = 17$ **b** $r = 2.5$ and $h = 12$.

3 The sum, S, of the angles in a polygon with n sides is given by the formula $S = 180(n - 2)°$.

Find the sum of the angles, S, when:

a $n = 7$ **b** $n = 12$.

4 $V = u + ft$.

Find V when:

a $u = 40, f = 32$ and $t = 5$ **b** $u = 12, f = 13$ and $t = 10$.

5 $D = \dfrac{M}{V}$.

Find D when:

a $M = 28$ and $V = 4$ **b** $M = 8$ and $V = 5$.

 6 An entertainer charges £75 for every show he performs, plus an extra £50 per hour spent on stage.

The formula for calculating his charge is $C = 50t + 75$, where C is the charge in pounds and t is the length of the show in hours.

How much does he charge for a show lasting:

a 1 hour   **b** 3 hours   **c** 2 hours?

7 The area, A, of the triangle shown is given by the formula $A = \frac{1}{2} bh$.

Find the area, A, when:

a $h = 12$ cm and $b = 5$ cm

b $h = 9$ cm and $b = 8$ cm.

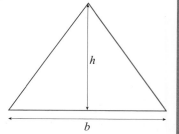

8 The formula $F = 2C + 30$ approximately converts temperatures in degrees Celsius (C) to degrees Fahrenheit (F).

Convert each of these temperatures to degrees Fahrenheit using this formula.

a 45 °C   **b** 40 °C   **c** 65 °C   **d** 100 °C

Investigation: Values less than 1

A What values of n can be substituted into n^2 to give n^2 a value less than 1?

B What values of n can be substituted into $(n - 4)^2$ that give $(n - 4)^2$ a value less than 1?

C What values of n can be substituted into $\dfrac{1}{n}$ that give $\dfrac{1}{n}$ a value less than 1?

D Find at least five different expressions in x that give the value 10 when $x = 2$ is substituted into them.

Ready to progress?

I can solve simple equations.

I can solve simple equations involving brackets and fractions.
I can substitute values into formulae.

Review questions

1 Solve each equation.

a $x + 2 = 8$	**b** $x - 3 = 8$	**c** $x + 5 = 9$	**d** $x - 4 = 6$
e $3m = 15$	**f** $5m = 35$	**g** $6m = 48$	**h** $7m = 21$
i $\dfrac{x}{2} = 8$	**j** $\dfrac{x}{4} = 2$	**k** $\dfrac{x}{3} = 9$	**l** $\dfrac{x}{4} = 5$
m $k + 14 = 6$	**n** $k + 8 = 5$	**o** $m + 7 = 3$	**p** $x + 8 = 1$

2 Solve each equation.

a $2(m + 3) = 12$	**b** $3(t - 5) = 21$	**c** $3(y - 3) = 18$	**d** $5(x + 3) = 30$
e $2(n + 7) = 16$	**f** $2(q - 4) = 8$	**g** $4(p - 4) = 16$	**h** $6(t - 4) = 36$
i $4(q - 2) = 20$	**j** $2(g + 3) = 12$	**k** $6(y - 4) = 36$	**l** $2(m - 4) = 16$
m $3(t - 6) = 27$	**n** $8(x + 3) = 40$	**o** $3(n + 5) = 21$	**p** $3(p - 2) = 36$

3
a Write down an expression for the area of this rectangle.

b The area of the rectangle is 52 cm².

Write down an equation involving x.

c Solve the equation.

d The perimeter of the rectangle is given by the formula $P = 2(t + 4)$

The perimeter of the rectangle is 22 cm.

Write down an equation involving x.

e Solve the equation.

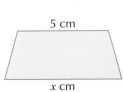

t cm

4 cm

4 The area of this trapezium is calculated using:

area = $3(5 + x)$

a i Write an equation involving x when the area is 36 cm².

ii Solve this equation.

b The area of the trapezium is now 42 cm².

i Write an equation involving x.

ii Solve this equation.

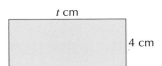

5 cm

x cm

5 For each of these number walls:

   i   use the number at the top to write an equation involving x

   ii solve the equation.

 Hint Remember that the number in each brick is the sum of the numbers in the two bricks below it.

a
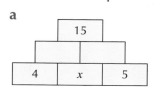

b
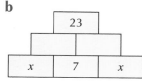

c
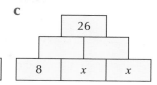

6 A window cleaner charges £3 for every visit to a house, plus an extra £2 for every window cleaned.

The formula for calculating his charge is $C = 2w + 3$, where C is the charge in pounds and w is the number of windows cleaned.

  a How much does he charge for visiting a house and cleaning these numbers of windows?

     i   4 windows   **ii**   5 windows   **iii**   8 windows

  b The window cleaner went to a house and charged £27.

     i   Write an equation involving w from this information.

     ii Solve the equation to find how many windows were cleaned at this house.

7 This is a page from Vicky's exercise book.

1. $5x + 2 = 22$
      $5x = 20$
       $x = 5$

2. $\dfrac{x}{3} = 12$
      $x = 4$

3. $3(t - 4) = 18$
     $t - 4 = 6$
       $t = 2$

4. $\dfrac{x}{4} + 7 = 10$
      $\dfrac{x}{4} = 17$
      $x = 61$

Vicky has made a mistake in each question.

Explain what she has done wrong in each one and correct it for her.

Reasoning
Old trees

Until 2013, the oldest individual tree in the world was Methuselah, a 4845-year-old Great Basin bristlecone pine in the White Mountains of California.

But in 2014 a tree was found to be a 5062-year-old pine also in the White Mountains. So this is now the oldest known living tree in the world.

There are a number of ways to find out how old a tree is.

Here are five:

a know when it was planted

b Use the tree ageing formula

Recognise the type of tree and measure its diameter in inches at your chest height. Then calculate the age of the tree by the rule:

tree's age = diameter × growth factor.

Some known growth factors

maple: 3.0   oak: 5.0
American elm: 4.0   dogwood: 7.0

c Cut it down and count the rings.

d Count the number of whorls:

age = 4 × whorls + 1

whorls

e i Measure the circumference of the trunk at your chest height. Call this C.

ii Cut off a branch, measure it's circumference and call this B.

iii Count the number of rings in the branch R.

iv tree age $= \dfrac{R \times C}{B}$

1 When did the oldest living tree in the world start to grow?

2 Tom cut down a tree and counted 32 rings. How old was the tree?

3 Joseph wanted to find out how old a tree was in the garden of a house he had just bought. He counted 17 whorls on the tree. How old was the tree?

4 Hannah said a tree in the park was 100 years old. Explain how you could check this without cutting anything off it.

5 David planted a tree on his wedding day, 26 April 1985.
   How old will the tree be on:

    Hint Give your answer to the nearest year or similar.

   a 12 February 2014   **b** 20 August 2015?

6 James wanted to see how old a tree was in his garden, so he cut off a small branch, measured its circumference as 8 cm and counted seven rings in the branch.
   He then measured the circumference of the tree at his chest height as 72 cm.
   How old was the tree?

7 Jess found that the oak in the park had a diameter of 19 inches. **Hint** 1 inch = 2.5 cm.
   Approximately how old was the oak tree?

8 Joy and Chris were talking about a tree in the botanical gardens.

   Explain how they could settle their argument.
   Remember, you cannot cut anything off a tree in the botanical gardens.

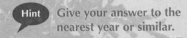
That Dogwood tree looks like it's more than 200 years old.

No, it's less than that but certainly more than 100 years old.

16

Comparing data

This chapter is going to show you:

- how to construct grouped frequency tables for data
- how to calculate a mean for discrete data
- how to construct frequency diagrams for discrete data
- how to compare two distributions by using an average and the range.

You should already know:

- how to construct a frequency table for discrete data
- how to find the mode, median and range for discrete data.

About this chapter

How can a coach select the best members for any sporting team, whether for the Olympics or a local football club? It helps to know and compare the records of all the applicants – not only their best times, points or scores over the previous few months but also how consistent their performances are.

You can use the same techniques to help you find the place that has the best record for high temperatures and sunshine for a good beach holiday. This chapter shows you how you can use data in this way to help make decisions.

16.1 Frequency tables

Learning objective

- To create a grouped frequency table from raw data

Key words

class	frequency
frequency table	grouped frequency table

You can draw up a **frequency table** to record how many times each value in a set of data occurs. The number of times a data value occurs is its **frequency**.

Often, data has a wide range of possible values; examples include masses or heights of all the pupils in a class. You have to group it together, to see any pattern.

In a **grouped frequency table**, you arrange information into **classes** or groups of data to do this. You can create frequency diagrams, from grouped frequency tables, to illustrate the data.

Example 1

These are the waiting times, in minutes, of people at a bus stop.

2, 10, 5, 1, 8, 9, 7, 6, 13, 5, 1, 9, 8, 7, 2, 1, 13, 0, 0, 2, 11, 8, 7, 5, 10, 6, 5, 14

Construct a grouped frequency table to represent the data.

Use a class size of 3, i.e. 0–2, 3–5, 6–8, 9–11, 12–14.

Record each time in the correct group using tallies.

Count the tallies and record the total in the frequency column.

Add up the frequencies and record the total in the table.

There are eight people who waited 0, 1 or 2 minutes.

There are four people who waited 3, 4 or 5 minutes.

There are eight people who waited 6, 7 or 8 minutes.

There are five people who waited 9, 10 or 11 minutes.

There are three people who waited 12, 13 or 14 minutes.

Here is the completed table.

Waiting time (minutes)	Tally	Frequency
0–2	卌 III	8
3–5	IIII	4
6–8	卌 III	8
9–11	卌	5
12–14	III	3
	Total	28

1. The table shows the lengths of time 28 customers spend in a shop.

Time (minutes)	Number of customers
0–10	13
11–20	8
21–30	7

Two more customers enter the shop. The first customer is in the shop for 5 minutes and the second customer is in the shop for 21 minutes.

Copy and update the table to include these two customers.

2. These are the heights (in centimetres) of 25 trees in an orchard.

273, 264, 226, 289, 276, 252, 284, 265, 238, 261, 299, 298, 264, 275, 307, 293, 229, 274, 255, 287, 286, 275, 302, 273, 288

a Copy and complete the grouped frequency table.

Height (centimetres)	Tally	Number of trees
220–239		
240–259		
260–279		
280–299		
300–319		

b What is the modal class?

 3. The masses (in kilograms) of fish caught in one morning by an angler are shown below:

1, 5, 3, 4, 6, 1, 2, 4, 5, 1, 5, 4, 3, 3

a Copy and complete the frequency table.

Mass (kilograms)	Tally	Number of fish
0–2		
3–5		
6–8		

b Which class contains the least fish?

c In the afternoon, the angler only caught four more fish, but this changed the class that had the least fish caught. Suggest four possible masses that the fish could have been. Explain why you chose these.

4 The midday temperature (in °C) of 20 ports visited on a cruise were recorded as:

17, 15, 26, 14, 18, 25, 17, 19, 22, 21, 23, 15, 17, 16, 20, 22, 20, 17, 18, 23.

a Copy and complete the frequency table.

Temperature (°C)	Tally	Frequency
13–15		
16–18		
19–21		
22–24		
25–27		

b What is the modal class?

5 For one month in a year Ian kept a record of how many customers came into his newspaper shop each day.

53, 42, 59, 46, 55, 54, 62, 50, 39, 54, 58, 47, 53, 56, 43, 62, 66, 55, 47, 60, 48, 52, 49, 56, 45, 65, 52, 60

a Using class sizes of 5, work out the modal class of the number of people that came into the shop.

b Explain how you can find out what month Ian did his survey.

Activity: Diagrams in this book

Paul says 'most pages in this book have either 2 or 3 diagrams'.

Investigate this statement to see if 2 or 3 diagrams is the most common class for the number of diagrams per page in the book.

16.2 The mean

Learning objective

- To understand and calculate the mean average of data

Key words	
average	mean
mean average	

The **mean** is the most commonly used average. It is also called the **mean average** or simply the **average**. The mean can be used only with numerical data.

The mean of a set of values is the sum of all the values divided by the number of values in the set.

$$\text{Mean} = \frac{\text{sum of all values}}{\text{number of values}}$$

The mean is a useful statistic because it takes all of the values into account, but it can be distorted by an extreme value. This is a value, in the set of data, that is much larger or much smaller than the rest. When there is an extreme value, the median is often used instead of the mean.

Example 2

Find the mean of 2, 7, 9, 10.

The mean is $\dfrac{2+7+9+10}{4} = \dfrac{28}{4} = 7$.

For more complex data, you can use a calculator. When the answer is not exact, the mean is usually given to one decimal place (1 dp).

Example 3

The ages of seven people are 40, 37, 34, 42, 45, 39, 35. Calculate their mean age.

The mean age is $\dfrac{40+37+34+42+45+39+35}{7} = \dfrac{272}{7} = 38.9$ (1dp).

Exercise 16B

1 Complete each calculation.

a The mean of 3, 8, 10 is $\dfrac{3+8+10}{3} = \dfrac{\square}{3} = \square$.

b The mean of 1, 4, 5, 6 is $\dfrac{1+4+5+6}{4} = \dfrac{\square}{4} = \square$.

c The mean of 2, 5, 9, 12 is $\dfrac{2+5+9+12}{4} = \dfrac{\square}{4} = \square$.

d The mean of 1, 1, 2, 5, 6 is $\dfrac{1+1+2+5+6}{5} = \dfrac{\square}{5} = \square$.

2 Complete each calculation.

a The mean of 4, 6, 11 is $\dfrac{4+6+11}{3} = \dfrac{\square}{\square} = \square$.

b The mean of 0, 2, 2, 6 is $\dfrac{0+2+2+6}{4} = \dfrac{\square}{\square} = \square$.

c The mean of 3, 3, 4, 4, 6 is $\dfrac{3+3+4+4+6}{5} = \dfrac{\square}{\square} = \square$.

d The mean of 1, 1, 3, 4, 11 is $\dfrac{1+1+3+4+11}{5} = \dfrac{\square}{\square} = \square$.

3 Complete each calculation.

a The mean of 0, 4, 5 is $\dfrac{0+4+5}{\square} = \dfrac{\square}{\square} = \square$.

b The mean of 1, 2, 8, 9 is $\dfrac{1+2+8+9}{\square} = \dfrac{\square}{\square} = \square$.

c The mean of 1, 1, 2, 5, 6 is $\dfrac{1+1+2+5+6}{\square} = \dfrac{\square}{\square} = \square$.

d The mean of 3, 7, 10, 20 is $\dfrac{3+7+10+20}{\square} = \dfrac{\square}{\square} = \square$.

4 Copy and complete each calculation.

a The mean of 2, 6, 7 is $\dfrac{\square+\square+\square}{\square} = \dfrac{\square}{\square} = \square$.

b The mean of 4, 7, 9, 16 is $\dfrac{\square+\square+\square+\square}{\square} = \dfrac{\square}{\square} = \square$.

c The mean of 1, 3, 10, 12, 14 is $\dfrac{\square+\square+\square+\square+\square}{\square} = \dfrac{\square}{\square} = \square$.

d The mean of 2, 4, 9, 9 is $\dfrac{\square+\square+\square+\square}{\square} = \dfrac{\square}{\square} = \square$.

5 Find the mean of each set of data.

a 7, 6, 5, 9, 3

b 12, 21, 29, 26, 18, 14

c 10, 11, 8, 25, 13, 16, 15

d 4, 6, 2, 8, 0, 5, 5, 2

6 Calculate the mean of each set of data, giving your answer to 1 dp.

a 7, 6, 4, 2, 3

b 2, 5, 7, 1, 8

c 6, 2, 2, 5, 7, 7, 7, 8

d 10, 7, 10, 8, 11

7 These are the heights, in centimetres, of 10 boys.

131, 146, 142, 135, 134, 145, 152, 131, 136, 148

a Calculate the mean height of the boys.

b What is the median height of the boys?

c What is the modal height of the boys?

d Which average do you think is the best one to use?
Explain your answer.

8 These are the numbers of children in the families on the same street as Sophia.

0, 0, 0, 0, 1, 1, 2, 2, 3, 3

a What is the mode?   **b** What is the median?   **c** What is the mean?

9 These are the shoe sizes of all of Joy's cousins.

3, 3, 3, 3, 3, 4, 5, 6, 6, 9, 10

a What is the mode?   **b** What is the median?   **c** What is the mean?

FS **10** These are the different prices of coffee on sale at a café.

£1.50, £1.70, £1.45, £1.60, £1.90, £1.65, £1.50, £1.50, £1.60, £1.90

The mean price of the coffees in the coffee shop down the road was £1.70.

Is the mean price of the coffee shop higher or lower than the cafe?

Explain your answer.

16.3 Drawing frequency diagrams

Learning objective

• To be able to draw a diagram from a frequency table

How could you record the different ways in which pupils travel to school?

When you have collected data from a survey, you can display it using different diagrams to make it easier to understand.

Bar charts can be used to show different categories, such as walking to school, using the school bus or going to school by taxi. This type of bar chart has gaps between the bars.

Bar charts can also be used for grouped data. For example, when recording word lengths for 100 words, you could group the number of letters per word into classes, such as 1–3, 4–6, 7–9 and 10–12. For bar charts like this, there are no gaps between the bars.

Bar charts are also used to show single values. A typical case would be, when rolling a dice 50 times, record the number of times that each score (1, 2, 3, 4, 5 or 6) is obtained.

Example 4

Construct a bar chart for the following data about the ways in which pupils travel to school.

How pupils travel to school	Number of pupils
Walk	4
Bus	5
Car	10
Cycle	6

Draw the bar chart as shown. The bar chart should have:

• a title

• labels on both axes

• gaps between the bars because the data is in categories.

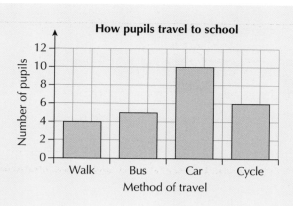

How pupils travel to school

Example 5

Jenna rolls a dice 50 times. Construct a bar-line graph to show her results.

Score on dice	Frequency
1	8
2	7
3	7
4	9
5	11
6	8

The length of each bar represents each frequency.

Draw the bar-line graph as shown. The bar-line graph should have:

- a title
- labels on both axes
- gaps between the bars because the data has single values.

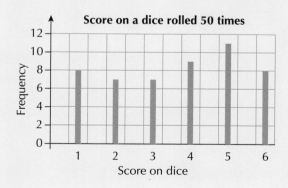

Score on a dice rolled 50 times

Exercise 16C

1 **a** Draw a bar chart to show the birthday season of a class of Year 8 pupils.

Birthday	Frequency
Spring	3
Summer	6
Autumn	8
Winter	7

b What is the mode season?

2 **a** Draw a bar chart to show the favourite field event of competitors.

Field event	Frequency
Javelin	12
Discus	6
Shot	8
Hammer	9

b What is the modal field event?

3 **a** Draw a bar-line graph to show the scores out of 5 in a test.

Score in test	Frequency
0	9
1	12
2	6
3	3
4	7
5	1

b What is the mode score?

4 **a** Use the data in the frequency table to draw a bar chart to show the number of words in 50 sentences.

Number of words	Frequency
0–10	8
11–20	12
21–30	25
31–40	5

b What is the modal class?

MR **5** **a** Explain why you cannot find a median for the data in question 1.

b Explain why you cannot find a median for the data in question 2.

c Explain how you can find the median for the data in question 3.

d Explain why you cannot find the median for the data in question 4.

6 The dual bar graph shows the mean monthly temperature for two cities.

   a Which city has the highest mean monthly temperature?

   b Which city has the lowest mean monthly temperature?

   c How many months of the year is the temperature higher in City A than City B?

   d What is the difference in mean temperature between the two cities in February?

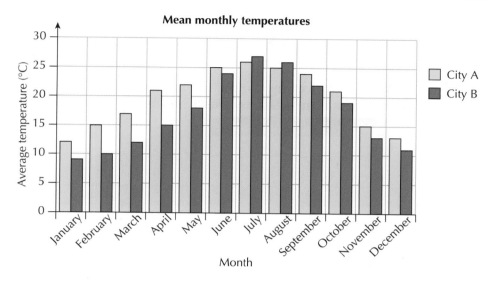

Activity: Comparing temperatures

Using the internet or an other source, compare the temperatures of two European resorts.

Make a poster to advertise one resort as being better than the other.

16.4 Comparing data

Learning objective

* To use the mean and range to compare data from two sources

The picture shows several shots made by a golfer. What is the range of the shots?

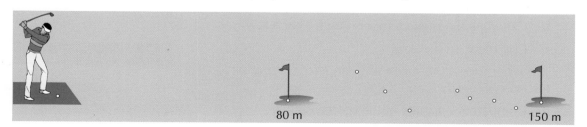

It is often important to know the range of results as this can show how consistent they are. A golfer whose shots ranged over 20 m to 150 m would be performing less consistently than this one.

Example 6

The table shows the median and range of basketball scores for two teams.

Compare the mean and range. What do they tell you?

	Team A	Team B
Mean	75	84
Range	20	10

The mean tells you that the average score for Team B is higher than that for Team A. So Team B has higher scores generally.

The range is the difference between their lowest and highest scores. As this is higher for Team A, there is a greater variation in Team A's scores. That is, Team A is less consistent than Team B.

Exercise 16D

1   The temperature of melting ice is 0 °C and the temperature of boiling water is 100 °C. What is the range of the two temperatures?

2   The times of four pupils in a 100-metre race are recorded as:

14.2 s, 13.8 s, 15.1 s, 17.3 s

Write down the range of the times.

3   A factory manager records the start and finish times of a series of jobs as shown below.

Job number	1	2	3	4	5
Start time	8.00 am	8.20 am	8.50 am	9.15 am	9.30 am
Finish time	8.30 am	9.15 am	9.20 am	9.40 am	9.35 am

Work out the range of the time taken for each job.

 4   The minimum and maximum temperatures are recorded for five towns in March as below.

Town	Exeter	Sheffield	Pembroke	Bath	Newcastle
Minimum	4 °C	3 °C	2 °C	5 °C	1 °C
Maximum	13 °C	12 °C	16 °C	14 °C	10 °C

a   Find the range of the temperatures for each town.

b   Calculate the mean temperature for each town.

c   Comment on the data.

	Chris	Kath
Mean	54	61
Range	25	13

 5   The table shows the mean and range of a set of test scores for Chris and Kath.

a   Compare the means of the scores, stating who was better.

b   Compare the ranges, stating who was more consistent.

MR **6** The table shows the means and the ranges of the scores of Kyle and Lisa after playing on a computer game.

	Kyle	Lisa
Mean	875	1500
Range	200	650

    a Compare the means of their scores, stating who you think is the better player.

    b Compare the ranges, stating who is more consistent.

MR **7** The table shows the mode and range of shoe sizes for men and women in a high-street shop.

Compare the modes and the ranges. What do they tell you?

	Men	Women
Mode	9	6
Range	8	7

PS **8** Belinda bought three bottles of face wash. The total price was £18 and the range was £2.50. The cheapest bottle cost £4.50.

    a What was the cost of the most expensive bottle?

    b What was the cost of the other bottle?

Activity: Comparing populations

Use the internet to compare the population and area of China with the population and area of the United States of America.

16.5 Which average to use?

Learning objective

• To understand when each different type of average is most useful

How would you work out an average height for this group of people?

You cannot use the mode because all their heights are different, so none is more common than the rest.

You can use the median – note that this doesn't take into account how small the smallest value is.

The best type of average to use is the mean, which takes all the heights into account. You find this by adding all the values and dividing by the number of values.

There is a big difference between the shortest and tallest people. Therefore you could also use the range of heights to give some idea of how close to the average each person might be.

The table will help you decide which type of average to use for a set of data.

Average	Advantages	Disadvantages	Example
Mean	Uses all the data. Probably the most used average.	May not be representative when the data contains an extreme value.	1, 1, 1, 2, 4, 15 Mean = $\dfrac{1+1+1+2+4+15}{6} = 4$ which is a higher value than most of the data.
Median	Only looks at the middle values, so it is a better average to use when the data contains extreme values.	Not all values are used, so could be misleading.	1, 1, 3, 5, 10, 15, 20 Median = 4th value = 5 Note that the median is close to the values 1, 1 and 3 but further from the values 10, 15 and 20.
Mode	It is the most common value. Can be used for non-numerical data.	When the mode is an extreme value, it is misleading to use it as an average. May not exist.	Weekly wages of a boss and his four staff: £150, £150, £150, £150, £1000. Mode is £150 but mean is £320.
Range	It measures how spread out the data is.	It only looks at the two extreme values, which may not represent the spread of the rest of the data.	1, 2, 5, 7, 9, 40. Range 40 − 1 = 39 Without the last value (40), the range would be only 8.

Exercise 16E

1 The time (in seconds) to complete a short task is recorded for each of 15 pupils:
10, 10, 10, 10, 11, 11, 12, 12, 12, 13, 14, 15, 15, 16, 17
The values are then grouped into a frequency table.

a Write down the mode.

b The median is 12. Explain why the median is more useful than the mode in this case.

Time (seconds)	Frequency
10	4
11	2
12	3
13	1
14	1
15	2
16	1
17	1

2 Look at each set of data and the average which has been calculated. Give one reason why the average stated may not be the best average to use.

a 3, 3, 5, 7, 8, 10; mode = 3   **b** 0, 1, 2, 2, 8, 14, 16; mode = 2

c 1, 4, 7, 8, 20, 21, 32; median = 8   **d** 2, 3, 6, 7, 10, 10, 10; mode = 10

e 2, 2, 2, 2, 14, 16, 28; median = 2   **f** 0, 1, 4, 6, 9, 100; mean = 20

3 Look at each set of data and calculate the two averages given. State which average, if any, is best suited to that particular set.

    a   1, 2, 4, 7, 12, 100; median and mean

    b   9, 10, 10, 10, 91; mode and mean

    c   1, 100, 101, 102, 106; median and mean

    d   1, 3, 5, 6, 7, 8; median and mean

    e   1, 1, 1, 7, 10, 10, 10; median and mode

    f   2, 5, 8, 10, 15; median or mean

 4 These are the times (in seconds) that some people took to say the alphabet.

10, 11, 11, 12, 12, 12, 13, 14, 14, 14, 15, 15, 15, 16, 17

    a   Copy and complete the frequency table and find the modal class.

    b   Explain why the mode is unsuitable for the ungrouped data, but the modal class is suitable for the grouped data.

Time, T (seconds)	Tally	Frequency
10–11		
12–13		
14–15		
16–17		

 5 A factory employs 50 people. Of these, 25 are workers who each earn less than £350 a week, 20 are apprentices earning less than £100 a week and 5 are managers earning more than £600 a week each.

    a   Which average would you use to argue that pay at the factory was low?

    b   Which average would you use to argue that pay at the factory was reasonable?

    c   In discussions about average pay, which average would be used by:

        i   the workers   **ii**   the managers   **iii**   the apprentices?

        Give reasons for your answers.

Activity: Football match attendances

Collect the attendances at English Premiership football matches over one weekend. Calculate the range of this set of data.

Repeat this exercise for the Scottish Premier division.

Compare the differences in the distributions of the data. Explain why the range is probably more suitable for the English division than the Scottish division.

Repeat the calculations but ignore the largest attendance in each division. What effect does this have on your answers?

Ready to progress?

I can construct and interpret bar charts.

I can find the mean for a set of data.
I can compare the range and the mean from two sets of data.
I can decide which average is the best to use in different circumstances.

Review questions

1 a Draw a bar chart to show the colours of cars in a car park.

Colour	Frequency
White	2
Blue	10
Silver	11
Red	3

b What is the modal colour?

2 A group of friends were talking about how many devices they had in their homes where they could watch TV.

These are the numbers of devices.

6, 3, 3, 9, 8, 11, 4

a What is the mode?
b How many friends were in this group?
c What is the median number of devices?
d Callum, another friend arrives. They ask him how many devices there are in his home.

The median now becomes 5.

Suggest how many devices there are in Callum's home. Give reasons for your answer.

3 The table shows the mean and range of a set of test scores for Pete and Rodrigo.

Pupil	Pete	Rodrigo
Mean	44	51
Range	15	3

a Compare the means of the scores, stating who was better.
b Compare the ranges, stating who was more consistent.

4 These are the average temperatures (in °C) of 20 English resorts one summer.

16, 14, 13, 17, 16, 18, 21, 20, 22, 14, 18, 15, 19, 19, 16, 17, 19, 21, 16, 23

    a Copy and complete the frequency table.

Temperature (°C)	Tally	Frequency
13–15		
16–18		
19–21		
22–24		

    b Which is the modal class?

5 **a** Tom played four games in a competition.

    In three games, Tom scored 12 points each time. In the other game he scored no points. What was Tom's mean score over the four games?

    b Helen only played two games.

    Her mean score was 6 points. Her range was 2 points.

    What points did Helen score in her two games?

    c Murray played three games.

    His mean score was also 6 points. His range was also 2 points.

    What points did Murray score in his three games?

6 Look at each set of data and calculate the two averages.

State which average, if any, is best suited to that particular set.

Give reasons for your choice.

    a 2, 3, 5, 8, 50; median and mean   **b** 10, 11, 11, 11, 57; mode and mean

    c 2, 200, 200, 202, 206; median and mean   **d** 2, 4, 6, 7, 8, 9; median and mean

    e 2, 2, 2, 8, 11, 11, 81; median and mode   **f** 1, 4, 7, 9, 14; median and mean

7 There are 720 pupils in a school.

104 of these pupils are left handed.

    a The pie chart is not drawn accurately.

    What should the angles be? Show your working.

    b Exactly a quarter of these 104 pupils are boys.

    From this information, is the percentage of boys in this school that are left handed 4%, 8%, 15%, 25%, 40% or is it not possible to tell?

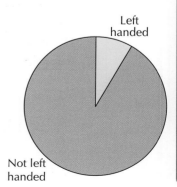

Left handed

Not left handed

Problem solving
Questionnaire

Chris created a questionnaire for his year group. He asked 25 boys and 25 girls the following questions.

a How many music downloaded tracks do you have?

b How many CDs do you possess?

c Which is your favourite Wii sporting game?

This is a summary of his results.

Boy/girl	Downloads	CDs	Wii game	Boy/girl	Downloads	CDs	Wii game
Boy	1824	12	Bowling	Girl	2012	17	Bowling
Girl	1632	17	Bowling	Boy	3008	32	Golf
Boy	2187	27	Boxing	Girl	3654	43	Bowling
Girl	2562	34	Bowling	Girl	2219	26	Golf
Girl	2193	32	Bowling	Boy	2187	32	Boxing
Boy	1086	29	Boxing	Boy	3280	44	Boxing
Boy	1243	43	Golf	Boy	3098	53	Boxing
Boy	1786	41	Boxing	Girl	2021	45	Bowling
Girl	1329	23	Bowling	Girl	783	36	Bowling
Girl	958	16	Golf	Boy	1086	28	Boxing
Boy	1982	26	Boxing	Boy	762	16	Tennis
Girl	1090	37	Bowling	Girl	542	19	Tennis
Girl	1824	24	Bowling	Boy	781	20	Golf
Boy	1655	16	Golf	Girl	2005	31	Bowling
Boy	2311	33	Boxing	Girl	1019	40	Bowling
Girl	4044	45	Bowling	Boy	2109	43	Bowling
Boy	4109	52	Tennis	Girl	3218	55	Golf
Girl	3381	37	Tennis	Girl	1894	23	Bowling
Girl	564	13	Bowling	Boy	2015	27	Baseball
Girl	2034	31	Bowling	Boy	2415	35	Boxing
Boy	4298	50	Golf	Girl	953	31	Golf
Boy	3026	43	Bowling	Girl	2180	28	Bowling
Boy	2051	25	Boxing	Boy	2876	46	Boxing
Girl	1980	30	Bowling	Boy	2133	34	Boxing
Boy	1950	17	Bowling	Girl	1066	26	Bowling

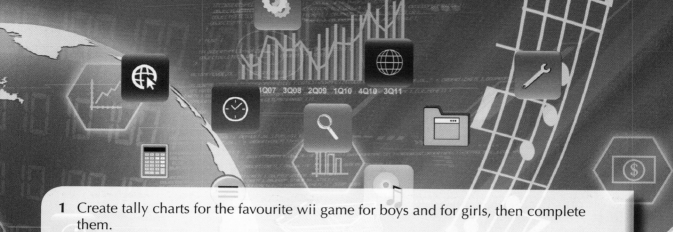

1 Create tally charts for the favourite wii game for boys and for girls, then complete them.

2 From your tally charts create pie charts for:

   a boys   b girls.

3 Create a grouped frequency table, using class sizes of 10, for the number of CDs for:

   a boys   b girls.

4 Create a frequency chart to help compare any differences between the number of CDs for girls and boys.

5 a Copy and complete the frequency table for the number of downloaded tracks for

     i girls   ii boys.

Class	Tally	Frequency
0–1000		
1001–2000		
2001–3000		
3001–4000		
4001–5000		

   b Draw a chart to illustrate the different numbers downloaded by boys and girls.

Glossary

algebraic expression An expression that contains numbers, variables and one or more arithmetic operations.

algebraic function machine A diagram that helps to identify the inputs, operations and outputs for an algebraic rule.

angle of rotation The angle through which an object is rotated, to form the image.

approximate Work out a value that is close but not exactly equal to another value, and can be used to give an idea of the size of the value; for example, a journey taking 58 minutes may be described as 'taking approximately an hour'; the sign ≈ means 'is approximately equal to'.

arc Part of the circumference of a circle.

average A central or typical value of a set of data, that can be used to represent the whole data set; mean, median and mode are all types of average.

base One of the sides of a 2D shape, usually the one drawn first, or shown at the bottom of the shape.

biased Not random, for example, attaching a small piece of sticky gum to the edge of a coin may cause it to land more frequently on Heads than on Tails: the coin would be biased.

brackets Symbols used to show expressions that must be treated as one term or number. Under the rules of BIDMAS, operations within brackets must be done first; for example: $2 \times (3 + 5) = 2 \times 8 = 16$ whereas $2 \times 3 + 5 = 6 + 5 = 11$.

centre The point inside the circle that is the same distance from every point on the circumference.

centre of rotation The point about which an object or shape is rotated.

chord A straight line joining two points on the circumference of a circle; a diameter is a chord that passes through the centre of the circle.

circumference The perimeter of a circle; every point on the circumference is the same distance from the centre, and this distance is the radius.

class A small range of values within a large set of data, treated as one group of values.

common factor A factor that divides exactly into two or more numbers; 2 is a common factor of 6, 8 and 10.

common multiple A multiple that appears in the times tables of two or more numbers.

compound shape A shape made from two or more simpler shapes; for example, a floor plan could be made from a square and a rectangle joined together.

congruent Exactly the same shape and size.

cube 1 In geometry, a 3D shape with six square faces, eight vertices and 12 edges.

2 In number and algebra, the result of multiplying a number or expression raised to the power of three: n^3 is read as 'n cubed' or 'n to the power of three': for example: 2^3 is the cube of 2 and $(2 \times 2 \times 2) = 8$.

cube root For a given number, a, the cube root is the number b, where $a = b^3$; for example, the cube root of 8 is 2 since $2^3 = 8$. The cube root of 8 is recorded as $\sqrt[3]{8} = 2$.

diameter A straight line joining two points on the circumference of a circle, and passing through the centre.

direct proportion   A relationship in which one variable increases or decreases at the same rate as another; in the formula $y = 12x$, x and y are in direct proportion.

direction of rotation   The direction, clockwise or anticlockwise, in which an object is rotated to form an image.

distance–time graph   A graph showing the distance travelled (vertical axis) against the time taken (horizontal axis), for a journey.

divisible   Able to be divided exactly; 6 is divisible by 2 and by 3, 8 is divisible by 2 and 4 but not by 3.

equally likely   When the probabilities of two or more outcomes are equal; for example, when a fair six-sided die is thrown, the outcomes 6 and 2 are equally likely with probabilities of $\frac{1}{6}$.

equation   A number sentence stating that two expressions or quantities are of equal value, for example, $x + 2y = 9$; an equation always contains an equals sign (=).

estimate   **1** State or guess a value, based on experience or what you already know.

2 A rough or approximate answer.

event   Something that happens, such as the toss of a coin, the throw of a dice or a football match.

expand   To expand a term with brackets, you multiply everything inside the brackets by the value in front of the brackets.

experimental probability   The probability found by trial or experiment; an estimate of the true probability.

factor   A number that divides exactly into another number, without leaving a remainder.

factor tree   A diagrammatic method of finding the prime factors of a number, by dividing it by its prime factors.

fair   The probability of each outcome is similar to the theoretical probability.

Fibonacci sequence   A sequence of numbers in which the third and subsequent terms are formed by adding the two previous terms: 1, 1, 2, 3, 5, 8, …

formula   A rule that connects two or more variables; a mathematical rule, using numbers and letters, that shows a relationship between variables; for example, the conversion formula from temperatures in Fahrenheit to temperatures in Celsius is: $C = \frac{5}{9}(F - 32)$.

frequency   The number of times a particular item appears in a set of data.

frequency table   A table showing data values, or ranges of data values, and the numbers of times that they occur in a survey or trial.

grouped frequency table   A frequency table in which the data is arranged in groups or ranges, to make it easier to work with.

graph   A diagram showing the relation between certain sets of numbers or quantities by means of a series of values or points plotted on a set of axes.

height   The vertical distance, from bottom to top, of a 2D or 3D shape.

highest common factor (HCF)   The largest number that is a factor common to two or more other numbers.

index   Power; in 3^4, 4 is the index.

index form   A number that is expressed as another number raised to a power is in index form.

integer   Whole number, may be positive or negative, including zero.

inverse operation   An operation that reverses the effect of another operation; for example, addition is the inverse of subtraction, division is the inverse of multiplication.

inverse proportion   A relationship between two variables in which as one value increases, the other decreases; in the formula $xy = 12$, x and y are in inverse proportion.

like terms   Terms in which the variables are identical, but have different coefficients; for example, $2ax$ and $5ax$ are like terms but $5xy$ and $7y$ are not. Like terms can be combined by adding their numerical coefficients so $2ax + 5ax = 7ax$.

linear equation   An equation such as $y = 4x - 7$, that will produce a straight-line graph.

lowest common multiple (LCM)   The lowest number that is a multiple of two or more numbers; 12 is the lowest common multiple of 2, 3, 4 and 6.

mean   An average value of a set of data, found by adding all the values and dividing the sum by the number of values in the set; for example, the mean of 5, 6, 14, 15 and 45 is $(5 + 6 + 14 + 15 + 45) \div 5 = 17$.

mean average   See mean.

multiple   A number that results from multiplying one number by another; multiples appear in the multiplication table of a number.

multiply out   To multiply out brackets, you multiply everything inside the brackets by the number or term outside the brackets.

negative number   A number that is less than zero.

nth term   An expression in terms of n; it allows you to find any term in a sequence, without having to use a term-to-term rule.

outcome   A result of a trial or event.

parallel   Lines that are always the same distance apart, however far they are extended.

parallelogram   A quadrilateral with two pairs of parallel sides; the opposite sides are equal in length.

percentage   A number written as a fraction with 100 parts, but instead of writing it as a fraction out of 100, you write the symbol % at the end, so $\frac{50}{100}$ is written as 50%.

perpendicular   At right angles to.

perpendicular height   The distance between the base of a 2D shape and its topmost point or vertex, measured at right angles to the base.

positive number   A number that is greater than zero.

power   How many times you use a number or expression in a calculation; it is written as a small, raised number; for example, 2^2 is 2 multiplied by itself, $2^2 = 2 \times 2 = 4$ and 4^3 is $4 \times 4 \times 4 = 64$.

power of 10   A number that can be expressed as 10 multiplied by itself one or more times, for example, 100, 1000.

prime factor   A factor that is a prime number; 2 and 3 are the prime factors of 6.

prime number   A number that has exactly two factors, itself and 1; the number 1 is not a prime number, as it only has one factor.

probability   The measure of how likely an outcome of an event is to occur. All probabilities have values in the range from 0 to 1.

probability scale   A scale or number line, from 0 to 1, sometimes labelled with impossible, unlikely, even chance, etc., to show the likelihood of an outcome of an event occurring. Possible outcomes may be marked along the scale as fractions or decimals.

proportional   In proportion.

quadratic   Containing a term with a squared variable, such as $y = 2x^2 + 4$.

radius   The shortest distance between the centre of a circle and its circumference.

ratio   A way of comparing the sizes of two or more numbers or quantities; for example, if there are five boys and ten girls in a group, the ratio of boys to girls is 5 : 10 or 1 : 2, the ratio of girls to boys is 2 : 1. The two numbers are separated by a colon (:).

right angle   One quarter of a complete turn. An angle of 90°.

rotation   A turn about a central point, called the centre of rotation.

round   Approximate according to a given condition, such as a number of decimal places or significant figures.

sample space   The set of all possible outcomes for an event or trial.

scale   The ratio of the length on the image to the length on the object.

scale diagram   A diagram that represents something much larger or much smaller, in which the lengths on the image are in direct proportion to the lengths on the object.

scaling   A method used in drawing statistical diagrams, such as pie charts; data values are multiplied or divided by the same number, so that they can be represented proportionally in a diagram.

scatter graph   A graphical representation showing whether there is a relationship between two sets of data.

sector   A region of a circle, like a slice of a pie, bounded by an arc and two radii.

semicircle   Half of a circle, based on a diameter.

significant   In a number, the digits that give an approximation of its value are significant.

significant figure   In the number 12 068, 1 is the first and most significant figure and 8 is the fifth and least significant figure. In 0.246 the first and most significant figure is 2. Zeros at the beginning of a number are not significant figures.

solve   To find the value or values of a variable (x) that satisfy a given equation.

square   Multiply a number by itself.

square root   For a given number, a, the square root is the number b, where $a = b^2$; for example, a square root of 25 is 5 since $5^2 = 25$. The square root of 25 is recorded as $\sqrt{25} = 5$. Note that a positive number has a negative square root, as well as a positive square root; for example, $(-5)^2 = 25$ so it is also true that $\sqrt{25} = -5$.

term   **1** A part of an expression, equation or formula. Terms are usually separated by + and − signs.

2 A number in a sequence or pattern.

total frequency   The result of adding together all of the frequencies in a data set.

translate   Move in a straight line, horizontally, vertically or diagonally.

translation   A movement along, up or diagonally on a coordinate grid.

unit fraction   A fraction with 1 as the numerator.

unitary method   A method of carrying out a calculation to find the value of a number of items by first finding the value of one of them.

variable   A quantity that may take many values.

Index

William Collins's dream of knowledge for all began with the publication of his first book in 1819. A self-educated mill worker, he not only enriched millions of lives, but also founded a flourishing publishing house. Today, staying true to this spirit, Collins books are packed with inspiration, innovation and practical expertise. They place you at the centre of a world of possibility and give you exactly what you need to explore it.

Collins. Freedom to teach.

Published by Collins
An imprint of HarperCollins*Publishers*
77–85 Fulham Palace Road
Hammersmith
London
W6 8JB

Browse the complete Collins catalogue at
www.collins.co.uk

ISBN-13 978-0-00-753774-7

The authors Kevin Evans, Keith Gordon, Chris Pearce, Trevor Senior and Brian Speed assert their moral rights to be identified as the authors of this work.

British Library Cataloguing in Publication Data
A catalogue record for this publication is available from the British Library.

Commissioned by Katie Sergeant
Project managed by Elektra Media Ltd
Developed and copy-edited by Jim Newall
Edited by Helen Marsden
Proofread by Amanda Dickson
Illustrations by Ann Paganuzzi, Nigel Jordan and Tony Wilkins
Typeset by Jouve India Private Limited
Indexing by Indexing Specialists (UK) Ltd
Cover design by Angela English

Printed and bound by L.E.G.O. S.p.A. Italy

Acknowledgements
The publishers wish to thank the following for permission to reproduce photographs. Every effort has been made to trace copyright holders and to obtain their permission for the use of copyright materials. The publishers will gladly receive any information enabling them to rectify any error or omission at the first opportunity.

(t = top, c = centre, b = bottom, r = right, l = left)

Cover gyn9037/Shutterstock, p 6 stocksolutions/Shutterstock, p 26–27 DOPhoto/Shutterstock, p 28 Brian A Jackson/Shutterstock, p 46–47 Hallgerd/Shutterstock, p 48 Paolo Nespoli/ESA/NASA via Getty Images, p 68–69 clearviewstock/Shutterstock, p 70 Lightspring/Shutterstock, p 80–81 bikeriderlondon/Shutterstock, p 82 Filip Fuxa/Shutterstock, p 94–95 Maceofoto/Shutterstock, p 96 jean morrison/Shutterstock, p 116–117 CandyBox Images/Shutterstock, p 118 design36/Shutterstock, p 134–135 BHJ/Shutterstock, p 136 Suppakij1017/Shutterstock, p 150–151 aquatic creature/Shutterstock, p 152 Sergey Nivens/Shutterstock, p 168–169 Adisa/Shutterstock, p 170 bekulnis/Shutterstock, p 186–187 Alyona Burchette/Shutterstock, p 188 *Mona Lisa*, c.1503-6 (oil on panel), Vinci, Leonardo da (1452-1519)/Louvre, Paris, France/Giraudon/The Bridgeman Art Library, p 202–203 Gordon Ball LRPS/Shutterstock, p 204 andersphoto/Shutterstock, p 218–219 pjmorley/Shutterstock, p 220 The *Proportions of the human figure (after Vitruvius)*, c.1492 (pen & ink on paper), Vinci, Leonardo da (1452-1519)/Galleria dell' Accademia, Venice, Italy/The Bridgeman Art Library, p 238–239 Taina Sohlman/Shutterstock, p 240 Art Konovalov/Shutterstock, p 250–251 ajt/Shutterstock, p 252 SergeyDV/Shutterstock, p 266–267 Darren J. Bradley/Shutterstock, p 266 Mega Pixel/Shutterstock, p 268 Robert Nyholm/Shutterstock, p 284–285 nmedia/Shutterstock.